Features of Liquid Crystal Display Materials and Processes

Features of Liquid Crystal Display Materials and Processes

Editor

Dong Chin Sun

scitus
academics

Features of Liquid Crystal Display Materials and Processes
Edited by **Dong Chin Sun**

ISBN: 978-1-68117-486-0
Library of Congress Control Number: 2016942838

SCITUS Academics LLC,
www.scitusacademics.com
Box No. 4766, 616 Corporate Way,
Suite 2, Valley Cottage,
NY 10989

Notice

Printed in United States of America on Acid Free Paper

Preface

Liquid Crystal Display (LCD) is an electric screen unit that operates at a level of liquid crystal using a different electrical voltage, thereby inducing changes in its optical qualities. LCDs are usually used for portable electronic video games, such as digital cameras and camcorders, video projection methods, electric billboards, computer monitors and flat-panel televisions. There are many components in a working LCD: glass screen, mechanical package, control electronics, drive electronics and power cord. The glass screen, between which the fluid crystals lie, is covered with column and row electrodes and now contains pads to connect the electronic drive (electric current) to each row and the electrode of the column. The electronic drive devices are integrated circuits that last supply both the row and the column electrodes with "drive". The control electronics is also integrated circuits. They decode and interpret the new signals – for example from the laptop computer– and send them using the drive electronics. The physical package is the frame that mounts the printed circuit boards for the drive and controls the display glass equipment as well.

This book includes theoretical modelling for nonlinear optics and optoelectronics, along with experimental methods, and new approaches and reasons for laser screen technology, unit engineering for semiconductors, biotechnology, medicine, etc. Liquid crystal displays (LCDs) contain liquid crystals that are generated by electricity. They are most often used to display one or more lines of alpha numeric information in a variety of devices: fax machines, notebook computer screens, call counters, medical instruments, lightweight compact disks and clocks, etc. Probably the most expensive and advanced style-efficient matrix displays-are also used as handheld color TV screens. They are highly popular for large, high-definition TVs. The innovative concept, the method and the facts described here are valuable for readers in fundamental research and in the marketplace.

T

Table of Contents

CHAPTER 1

Precautionary Measures Necessary in Liquid Crystal Display (LCD) Television Maintenance by Electronic Technicians

Ifeanyi Benedict Ohanu, Theresa Chinyere Ogbuanya

Department of Vocational Teacher Education, University of Nigeria, Nsukka, Nigeria

ABSTRACT

The study was carried out to determine precautionary measures necessary in liquid crystal display (LCD) television maintenance by electronic technician. The population for the study was 3209 consisting of 11 electronic teachers and 3198 electronic technicians. 11 electronic teachers and 155 electronic technicians selected by simple random sampling were used for the study. Questionnaire consisting of 57 items was used to elicit responses from electronic teachers and electronic technicians regarding the two research questions. Data collected were analyzed using mean while t-test for independent sample was used to test the two null hypotheses at 0.05 level of significance. Analysis revealed the important components of LCD television that may occasionally developed faults and important precautionary measures that should be taken in LCD television maintenance. It was concluded that electronic technicians should carry out maintenance of LCD televisions by observing the precautionary measures to be taken in maintenance of LCD televisions.

KEYWORDS

Electronic Technology; Liquid Crystal Display Television; Maintenance; Precautionary Measures; Electronic Technician

1. INTRODUCTION

Every appliance used by men especially electronic appliances requires maintenance for steady functionality, durability, and even aesthetics. Maintenance is the process of keeping equipment, tools, materials, or system in good working condition. It involves decision making and planning by the personnel. Maintenance of electronic appliances can be in various forms such as dusting, lubricating, replacing burnt/damaged components, repairing, testing, using items for the purpose they are designed and following procedure for the operation of each appliance. Different types of maintenance have been mapped out for the maintenance of appliances such as: preventive maintenance, predictive and corrective maintenance. Preventive maintenance is systematic detection and prevention of potential faults before they occur (Bayle, [1]).

According to Operations and Maintenance Best Practices [2] , preventive maintenance is actions performed on a timeor machine-run-based schedule that detect, preclude, or mitigate degradation of a component or system with the aim of sustaining or extending its useful life through controlling degradation to an acceptable level. Corrective maintenance can also be referred to as breakdown maintenance. It is the kind of maintenance given to equipment or facility when it cannot function as it is supposed to. It is a response given to an unanticipated problem or emergency [1] . On the other hand, predictive maintenance is the type of maintenance that is carried out on an item/equipment when there is a warning signal of breakdown. It involves arresting the situation promptly before there is any major breakdown that will demand corrective maintenance. According to Operations and Maintenance Best Practices [2] , predictive maintenance differs from preventive maintenance by basing maintenance need on the actual condition of the machine rather than on some preset schedule. It is measurement that detects the onset of system degradation (lower functional state), thereby allowing causal stressors to be eliminated or controlled prior to any significant deterioration in the component physical state. Predictive maintenance is usually given on either daily, weekly, bi-weekly or monthly bases. Maintenance is required to be carried out on most electronic gadgets as a result of their complex nature, fragile components used in their construction, miniaturized nature of some components of present-day electronic gadgets, low current/voltage used by electronic gadgets, and so on. Due to their complex and fragile nature, good precautionary measures must be taken in the cause of maintenance of electronic gadgets.

Precautionary measures are care taken to avoid damage to equipment. Innovations in the components used in construction of electronic gadgets require serious precautions to be taken in carrying out any maintenance operation or activities on them. This calls technicians who are familiar with the components for effective maintenance of the electronic gadgets. Technicians are trained formally in technical colleges while artisans are trained informally through apprenticeship systems available in the society. Apprenticeship system has been available in Nigeria long before the advent of western system of education. Apprenticeship is the process whereby the apprentice studies the processes involved in a particular job so as to acquire the required skills under the tutelage of his master for a period of six months or over, depending on the agreement that the apprentice reached with his master. It is also a process whereby students try to acquire skills outside school hour which will enable them to work and earn money to take care of their needs while they are still in school. According to Leading Learning Council and Skill [3] , apprenticeships are training programmes undertaken in the workplace, giving young people the opportunity to earn money while developing valuable skills and qualifications. It provide the opportunity for learners to "earn and learn" in a wide range of occupations and sectors.

Technicians also enroll in apprenticeship system to learn from their masters by imitation of what their master does. Apprentice-ship system is available in different trades such as blacksmithing, welding, electronic maintenance, electrical installation, and so on. According to Okoro [4] , the apprenticeship programme produces semiskilled craftsmen in auto-mechanics, electrical installation, radio and TV repair, carpentry and joinery, cabinet making, building construction and many more areas. Apprentice in electronic maintenance learn skills involved in maintenance of different kinds of electronic gadgets. However, due to innovation in the field of electronics, some of these apprentices limit themselves to learn about the maintenance of few electronic gadgets so as to become efficient in maintenance of such gadgets. Technicians are also trained in technical colleges through formal education. Technical colleges are concerned with production of technicians who are skilled in different fields of human endeavour. The types of technicians produced at technical college level include: electronic/electrical technicians, automobile/metalwork technicians, building/woodwork technicians, and so on. Training of technicians at technical college involves imparting both theoretical and practical knowledge to the students under the guidance of the teacher within the school environment. Students at technical colleges are exposed to different skilled areas, through which one can be trained to become a self-reliant technician. This type of education received at technical college level is known as technical education.

Federal Republic of Nigeria [5] defined technical and vocational education as a comprehensive term referring to those aspects of the educational process involving in addition to general education: the study of technologies and related sciences and the acquisition of practical skills, attitudes, understanding and knowledge relating to occupations in various sectors of economic and social life. Technical education is an aspect of education that leads to acquisition of practical and applied skills as well as basic scientific knowledge [4] . At technical college level, technical education is provided to impart technical/occupational skills in the areas of electricity, electronics, metalwork, woodwork, automobile, and so on. This implies that electronic technology is an aspect of technical education.

Electronic technology is the process of applying scientific knowledge in the design, selection of materials, construction, operation and maintenance of electronics [6]. It provides the skills, knowledge and attitudes necessary for effective employment in specific occupation. Electronic technicians produced at technical college level are those that offered courses in electronic technology. They are supposed to be skilled in different areas of electronic technology such as maintenance of electronic gadgets, designing of electronic circuits, as well as construction of some electronic gadgets/appliances.

A technician is a person who is trained formally in a particular trade. Electronic technicians are trained to meet the requirements of man in the ever-changing technological society. One of such prominent changes in the field of electronics is the production of more sophisticated electronic gadgets of which Liquid Crystal Display (LCD) television is one of them. These new and sophisticated gadgets are constructed with different components which are more fragile, complex, costly, and can perform faster than the components used in the construction of earlier developed gadgets. LCD electronic gadgets are new in the society and they are available in different electronic gadgets such as wrist watches, laptop computers, desktop computers, calculators, phones, televisions, and so on.

According to Theraja and Theraja [7] , a liquid crystal display is a material (usually, an organic compound) which flows like a liquid at room temperature but whose molecular structure has some properties normally associated with solids (examples of such components are: cholesteryl nonanoate and p-azoxyanisole). Liquid crystal “cell” consists of a thin layer (about 10 μm) of a liquid crystal sandwiched between two glass sheets with transparent electrodes deposited on their inside faces. The LCD materials are used in construction of LCD televisions.

LCD televisions are complex and contain so many fragile components such as the LCD screen, speakers, audio and video components, among others. These components on their own consist of different circuits which are made up of several electronic components like resistors, capacitors, inductors and ICs. The LCD television has the advantage of reduced weight, low power consumption, among others [7] . The advantage of reduced weight of LCD televisions can be attributed to the components that make up LCD televisions which are usually of little weight. The LCD televisions are continuously displacing the use of Cathode Ray Tube (CRT) televisions. The increased level of acceptance of LCD televisions has led to the use of LCD televisions in so many homes, hotels, banks, hospitals, offices, and guest houses. However, these LCD televisions develop fault(s) at one time or the other. This calls for electronic technicians who are skilled in maintenance of LCD televisions to provide remedy to such fault(s).

The kind of remedy given to LCD television fault depends on the type and nature of the problem identified. Some of the ways of solving LCD television faults have been identified by [8] . According to Wistrom [8] , if there is no power/image, the power cord should be checked to make sure that it is fully inserted into the outlet. You may want to test the outlet with another appliance to see if there is a problem with the outlet. Turn the power button on, if the picture looks extremely distorted, external devices may be interfering with the signal. Remove any appliances near the television, and see if it works better. Fluorescent lights can also interfere with the image, so you may move the television to a new location. In carrying out the different solutions to the faults of LCD television, a lot of precautionary measures should be taken.

Maintenance of LCD televisions requires a lot of precautionary measures. This is owing to the fact that some sections of the LCD television circuit contain alternating current and high voltage. As such, great care must be taken to avoid electrocution. The following precautionary measures should be taken when working on any LCD television. Don't work alone—in the event of an emergency, another person's presence may be essential. Always keep one hand in your pocket when anywhere around a powered line-connected or high voltage system. Wear rubber bottom shoes or sneakers. Don't wear any jewelry or other articles that could accidentally contact circuitry and conduct current, or get caught in moving parts. Set up your work area away from possible grounds that you may accidentally contact [9] . LCD television set should be unplugged from socket outlets after use. This will help to avoid burning of the fuse or other components during lightening. Goldwasser [9] stated that ideally, electronic equipment should be unplugged (both AC line and phone line) during electrical storms if possible. Modern TVs, VCRs, microwave ovens, and even stereo equipment are particularly susceptible to lightning and surge damage because some parts of the circuitry are always alive and therefore have a connection to the AC line.

Sony [10] identified the following precautions to be taken on LCD television: unplug the television and other connected equipment from the wall outlet before you begin cleaning your television; wipe the LCD screen gently with a soft cloth; stubborn stains may be removed with a cloth slightly moistened with a solution of mild soap and warm water; if using a chemically pre-treated cloth, follow the instruction provided on the package; never use strong solvents such as a thinner, alcohol or benzene for cleaning; do not plug in the television into the wall outlet until the moisture from cleaning has evaporated.

No matter the extent of precautions taken, LCD television is a system in use and it is bound to breakdown from time to time. Hence knowledgeable and skilled technicians need to effectively observe precautionary measures and carry out maintenance of LCD television. Such competent electronic technicians will help in ensuring continuous usage of LCD television by the society by offering prompt and effective maintenance services, and they will also be gainfully employed or self-employed. Employment no doubt is one of the most vital instruments needed to speed-off development and curb various societal menaces in Nigeria. Taking good precautions in maintenance of LCD televisions will make the technician involve in maintenance of LCD television to have a good relationship with his customers. This is because good precautionary measures avoid development of other faults to LCD television in the cause of trying to solve existing fault(s). Therefore, this study intends to find out precautionary measures necessary in Liquid Crystal Display (LCD) television maintenance by electronic technicians. Safety precautions are necessary for any meaningful maintenance activity.

2. THEORIES

Safety theory by Weick [11] stated that safety is a dynamic non-event. Non-events, by their nature, tend to be taken for granted, particularly in the face of continuous and compelling productive demands. Safety is invisible in the sense that safe outcomes do not deviate from the expected, and so there is nothing to capture the attention. If people see nothing, they presume that nothing is happening, and that nothing will continue to happen if they continue to act as before. But this is misleading because it takes a number of dynamic inputs to create stable outcomes (James, 1998). This theory is related to this work because it emphasized the need for safety consciousness in maintenance activities. Maintenance of LCD television requires safety in handling the components and parts.

Another theory by James [12] stated that an ideal safety culture is the "engine" that drives the system towards the goal of sustaining the maximum resistance towards its operational hazards, regardless of the leadership's personality or current commercial concerns. This theory is related to this work because it emphasized the need for safety consciousness in maintenance activities. Maintenance of LCD television requires safety in handling the components and parts.

Hudson [13] stated that a safe environment is allowing dangerous activities to take place successfully, which means without harm or damage. What this means is that safety is more than a passive and well-meaning notion, such as "Thou shall do no harm". Instead safety is something that has to be actively managed to allow profit or advantage to be gained. This theory is related to the present study because it emphasized performing dangerous activities successfully. Such dangerous activities which should be performed successfully in maintenance of LCD television include care in handling the screen and the fragile components.

3. PURPOSE OF THE STUDY

The major purpose of the study was to find out precautionary measures necessary in liquid crystal display (LCD) television maintenance by electronic technicians. Specifically, the study sought:

1) The important components of LCD television that may occasionally develop faults

2) The important precautionary measures that should be taken in LCD television maintenance.

4. RESEARCH QUESTIONS

The following research questions guide the study:

1) What are the important components of LCD television that may occasionally develop faults?

2) What are the important precautionary measures that should be taken in LCD television maintenance?

5. HYPOTHESES

The following hypotheses that guided the study were tested at 0.05 levels of significance:

Ho_1: there is no significant difference between the mean responses of electronic teachers and technicians on the important components of LCD television that may occasionally develop faults.

Ho_2: there is no significant difference between the mean responses of electronic teachers and technicians on the important pre-cautionary measures that should be taken in LCD television maintenance.

6. METHOD

6.1. Design of the Study

The design adopted for this study was a descriptive survey research design. Descriptive survey design according to Ezeji [14] , is concerned with specifying the properties of educational and other phenomena. The design is appropriate because it sought the representative views of electronic teachers and technicians on the Troubleshooting of Faults in Liquid Crystal Display (LCD) Televisions.

6.2. Population of the Study and Sampling Procedure

The population for the study was 3209 persons. This consists of 11 electronic teachers in 10 Government Technical Colleges in Anambra State as well as 3198 electronic technicians in Anambra State. This figure was obtained from Anambra State Post Primary School Management Board [15] and Anambra State Ministry of Commerce and Industry [16] . All the 11 electronic technology teachers in 10 government technical colleges in Anambra State were used for study as well as 155 electronic technicians selected by simple random sampling. According to Gall, Gall, and Borg [17] , simple random sampling is the selection from a population through the process that provides every sample of a given size an equal probability of being selected. The teachers and the electronic technicians have similar culture.

6.3. Procedure

The questionnaire was the instrument used for eliciting responses from the respondents. The questionnaire was divided into three sections namely section A and section B, which consisted of 57 structured items. Section sought information on the important components of LCD television that may occasionally develop faults. Section B sought information on the important precautionary measures that should be taken in LCD television maintenance. The questionnaire utilized a 4-point scale of the questionnaire items was rated as follows: Very Important (VI), Important (I), Moderately Important (MI), and Not Important (UI) for research the two research questions. Three experts, two from department of Vocational Teacher Education, University of Nigeria, Nsukka and one technician were requested to carry out face validation of the instrument. Their comments, suggestions and advice were used to modify some of the items. The reliability of the instrument was established by trial testing using electronic teachers and technicians in Enugu State. Cronbach Alpha was used to determine the internal consistency of the items in the instrument. A reliability coefficient of 0.89 was obtained.

The researcher administered and collects the instrument with the help of six research assistants. The research assistants were requested to administer the instrument to the respondents and collect them immediately after completion by the respondents. The respondents were required to provide information that represent their opinion out of the available response alternative in the questionnaire. The instrument was collected immediately after its administration, the same day. It took two weeks to administer and collect the instrument and 95% of the administered instrument was returned. Data collected for this study was analyzed using mean. t-test for independent samples was used to test the hypothesis at 0.05 level of significance. The response to the items was interpreted using table of real limit of numbers as follows: Very Important (VI) = 3.50 - 4.00; Important (I) = 2.50 - 3.49; Moderately Important (MI) = 1.50 - 2.49; Not Important (NI) = 0.50 - 1.49. Base on the above limits, any item that has a mean score of 3.50 - 4.00 was regarded as Very Important, 2.50 - 3.49 was regarded as Important, 1.50 - 2.49 was regarded as Moderately Important, 0.50 - 1.49 was regarded as Not Important.

7. RESULTS

7.1. Table 1

Data presented in Table 1 show that all the items had their mean value ranging from 3.46 to 3.54 which falls within the response categories of Important and Very Important. This indicates that the items 1, 2, 19, 21, 24, 25, 27, 28, and 29 are important components of LCD television with their mean scores ranging from 3.46 to 3.49. While items 3 to 18, 20, 22, 23, 26, and 30 to 32 are Very Important components of LCD television with mean scores ranging from 3.50 to 3.56. The standard deviation for the items ranges from 0.54 to 0.64. This implies that the Electronic technology teachers and the Electronic technicians are homogenous in their responses on the important components of LCD television.

The result also shows that there is a significant difference in the mean responses of Electronic Teachers and Electronic Technicians on items 1, 2, 3, 7, 10, 11, 17 to 23, 29, and 31. Since their significance level ranges from 0.004 to 0.047. Therefore, the null hypothesis of no significance difference at 0.05 level of Significance was rejected for those items.

The result also revealed that there was no significance difference in the mean responses of Electronic Teachers and Electronic Technicians on items 4, 5, 6, 8, 9, 12 to 16, 24 to 28, 30, and 32. Since their significance ranges from 0.055 to 0.547 which are more than 0.05 level of significance. Therefore, the hypothesis of no significance difference was not rejected at 0.05 level of significance for those items.

7.2. Table 2

Data presented In Table 2 shows that all the items had their mean ranging from 3.28 to 3.45 which falls within the response categories of Important. This indicates that all the items are the important precautionary measures that should be taken in LCD television maintenance. The standard deviation for the items ranges from 0.62 to 0.67. This implies that the Electronic Teachers and Electronic Technicians are homogenous in their responses on the important precautionary measures that should be taken in LCD television maintenance. The result also shows that there is a significant difference in the mean responses of Electronic Teachers and Electronic Technicians on items 37, 38, and 57. Since their significance level are 0.036, 0.036, and 0.001 respectively which are less than 0.05 level of significance. Therefore, the null hypothesis of no significance difference at 0.05 level of significance was rejected for those items.

Table 1. Mean ratings, standard deviation and t-test analysis of responses of electronic technology teachers and technicians on the important components of LCD television that may occasionally develop faults.

s/n	Important components of LCD television that may occasionally develop faults	$\overline{X}$	S.D	t-test	Sig. (2-tailed)	Decision
1	Capacitors	3.46	0.61	2.952	0.004	S
2	Inductors	3.49	0.62	2.486	0.014	S
3	Tuner	3.51	0.62	2.002	0.047	S
4	Integrated circuit	3.53	0.59	1.618	0.108	NS
5	Video and audio demodulators	3.51	0.63	1.407	0.161	NS
6	Transistors	3.54	0.60	1.080	0.282	NS
7	LCD screen	3.51	0.63	2.576	0.011	S
8	Speakers	3.50	0.64	1.875	0.063	NS
9	Deflection yoke	3.51	0.64	1.937	0.055	NS
10	Power button	3.52	0.62	2.069	0.040	S
11	Resistors	3.51	0.62	2.621	0.010	S
12	Connectors	3.54	0.57	1.754	0.081	NS
13	Diodes	3.51	0.59	1.516	0.132	NS
14	Ambient sensors	3.54	0.56	1.166	0.246	NS
15	IR receivers	3.50	0.59	1.448	0.150	NS
16	Carbon	3.51	0.59	1.482	0.140	NS
17	Sockets	3.51	0.63	2.542	0.012	S
18	Fuse	3.51	0.61	2.072	0.040	S
19	Coils	3.49	0.62	2.519	0.013	S
20	Photocoupler	3.50	0.59	2.041	0.043	S
21	Varistor	3.49	0.61	2.565	0.011	S
22	Filter	3.50	0.59	2.041	0.043	S
23	Deflection coils	3.51	0.61	2.633	0.009	S
24	Ferrite bead	3.47	0.63	1.249	0.213	NS
25	Vibrator crystal	3.47	0.62	1.270	0.206	NS
26	Ultrafast rectifiers	3.50	0.62	0.604	0.547	NS
27	Supply cord	3.49	0.58	1.442	0.151	NS
28	Light emitting diodes	3.49	0.61	0.785	0.433	NS
29	X/Y capacitors	3.49	0.59	2.007	0.046	S
30	Secondary protection fuses	3.50	0.59	0.865	0.388	NS
31	Hook	3.50	0.62	2.553	0.012	S
32	Remote controller	3.52	0.60	1.522	0.130	NS

Key: NS = Not Significant, S = Significant, df = Degree of freedom (156), t-cal = Calculated value of t-test using SPSS. S.D = Standard Deviation. $\overline{X}_1$ = Mean for electronic technicians, $\overline{X}_2$ = Mean for electronic teachers, n_1 = Number of electronic technicians, n_2 = Number of electronic teachers, level of significance = 0.05.

The result also revealed that there was no significance difference in the mean responses of Electronic Teachers and Electronic Technicians on items 33 to 36, and 39 to 56. Since their significance level ranges from 0.116 to 0.919 which are more than 0.05 level of significance. Therefore the null hypothesis of no significance difference was not rejected at 0.05 level of significance for those items.

8. DISCUSSION

The result of this study revealed that 32 items constitute the important components of LCD televisions as indicated by the responses of the respondents. The items are: capacitors, inductors, tuner, integrated circuit, video and audio demodulators, transistors, LCD screen, speakers, deflection yoke, power button, resistors, connectors, diodes, ambient sensors, IR receiver, carbon, sockets, fuse, coils, photocoupler, varistor, filter, deflection coils, ferrite bead, vibrator crystal, ultrafast rectifiers, supply cord, light emitting diodes, X/Y capacitors, secondary protection fuses, hook, and remote controller. This is in line with Au Optronics Corporation [18] , [9] and Lewis [19] . Au Optronics Corporation [18] identified the main components of LCD television as power board, tuner, IR board, remote controller, LCD panel, back cover, chassis, stand, and keypad board. Some of the components of LCD television identified by Goldwasser [9] include: horizontal deflection, vertical deflection, LCD screen, chroma, demodulator, sync separators, among others.

Table 2. Mean, standard deviation and t-test analysis of responses of electronic technology teachers and technicians on the important precautionary measures that should be taken in LCD television maintenance.

s/n	Important precautionary measures that should be taken in LCD television maintenance.	$\bar{X}$	S.D	t-cal	Sig. (2-tailed)	Decision
34	Unplug the TV and other connected equipment from the wall outlet before you begin cleaning your TV.	3.42	0.70	1.419	0.158	NS
35	Wipe the LCD screen gently with a soft cloth.	3.42	0.66	1.440	0.152	NS
36	Ensure TV securely mounted on a stable surface before you begin loosen the parts to present it from toppling over.	3.45	0.66	1.583	0.116	NS
37	Stubborn stains may be removed with a cloth slightly moistened with a solution of mild soap and warm water.	3.44	0.65	2.118	0.036	S
38	Do not plug in the TV into the wall outlet until the moisture from cleaning has evaporated.	3.45	0.66	2.116	0.036	S
39	If using a chemically pre-treated cloth in cleaning the TV, follow the instruction provided on the package.	3.41	0.66	0.857	0.393	NS
40	Strong solvents such as a thinner, alcohol or benzene should not be used for cleaning LCD TV.	3.42	0.63	0.987	0.325	NS
41	LCD TV screen should not be exposed to direct illumination or sunlight.	3.42	0.65	0.957	0.340	NS
42	Do not install LCD television in a room with reflective wall and floor materials.	3.41	0.64	0.884	0.378	NS
43	LCD television should not be moved from a cold area to a warm area.	3.41	0.64	0.376	0.708	NS
44	LCD television should not be exposed to rain or excessive moisture.	3.41	0.63	0.898	0.371	NS
45	When turning off the TV by detaching the power cable or DC power cord, wait for 6 seconds before attach the power cable or DC power cord for normal operation.	3.41	0.64	0.376	0.708	NS
46	Ventilation holes on the television cabinet should not be blocked.	3.37	0.64	0.725	0.470	NS
47	Don't work alone so that in the event of an emergency another person's presence may be essential.	3.39	0.63	0.290	0.836	NS
48	Always keep one hand in your pocket when anywhere around a powered line-connected or high voltage system.	3.37	0.62	−0.383	0.702	NS
49	Wear rubber bottom shoes or sneakers when working on LCD television.	3.41	0.63	−0.193	0.847	NS
50	Work area should be set-up away from possible grounds that you may accidentally contact.	3.35	0.64	0.102	0.919	NS
51	Don't wear any jewelry or other articles that could accidentally contact circuitry and conduct current, or get caught in moving parts.	3.35	0.63	0.103	0.918	NS
52	Use an isolation transformer if there is any chance of contacting line connected circuits.	3.34	0.64	−0.497	0.620	NS
53	If you must probe live, put electrical tape over all but the last 1/16" of the test probes to avoid the possibility of an accidental short which could cause damage to various components.	3.32	0.64	0.483	0.630	NS
54	Perform as many tests as possible with power off and the equipment unplugged.	3.31	0.66	0.412	0.681	NS
55	Avoid working when you are really tired.	3.29	0.66	0.321	0.749	NS
56	Unplug the TV set from socket outlet after use to avoid burning of fuse or other electronic components as a result of lightening.	3.28	0.67	0.260	0.795	NS
57	If the television clicks with power-on but no other action, check other fuses in the circuit and replace them accordingly.	3.39	0.62	−4.312	0.001	S

Key: NS = Not Significant, S = Significant, df = Degree of freedom, t-cal = Calculated value of t-test using SPSS, S.D = Standard Deviation, $\bar{X}_1$ = Mean for electronic technicians, $\bar{X}_2$ = Mean for electronic teachers, n_1 = Number of Electronic technicians, n_2 = Number of electronic teachers, level of significance = 0.05.

The result of the analysis of the hypothesis revealed that there was no significant difference in the responses of electronic technology teachers and electronic technicians on 17 important components of LCD televisions that occasionally develop faults. However, significant difference existed in the respondents responses on 15 items as can be seen from Table 1. The hypothesis of no significant difference was upheld for 17 items, but was rejected for 15 items (see Table 1). Majority of the respondents agreed that the identified components are the important components of LCD televisions that occasionally develop faults. The above findings supports [18] , who identified the main components of LCD television as power board, tuner, IR board, remote controller, LCD panel, back cover, chassis, stand, and keypad board. The findings are also consistent with Goldwasser [9] who identified some components of LCD television to include: horizontal deflection, vertical deflection, LCD screens, chroma, demodulator, sync separators, and so on.

The result of the study shows that the following 25 items are the important precautionary measures that should be taken in maintenance of LCD televisions: the items are as follows: wear cotton gloves while troubleshooting, to avoid electrical shock, unplug the TV and other connected equipment from the wall outlet before you begin cleaning your TV, wipe the LCD screen gently with a soft cloth, ensure TV is securely mounted on a stable surface before you begin loosen the parts to present it from toppling over, stubborn stains may be removed with a cloth slightly moistened with a solution of mild soap and warm water, do not plug in the TV into the wall outlet until the moisture from cleaning has evaporated, if using a chemically pre-treated cloth in cleaning the TV follow the instruction provided on the package, strong solvents such as a thinner, alcohol or benzene should not be used for cleaning LCD TV, LCD TV screen should not be exposed to direct illumination or sunlight, do not install LCD television in a room with reflective wall and floor materials, LCD television should not be moved from a cold area to a warm area, LCD television should not be exposed to rain or excessive moisture, when turning off the TV by detaching the power cable or DC power cord wait for 6 seconds before attach the power cable or DC power cord for normal operation, ventilation holes on the television cabinet should not be blocked, don't work alone so that in the event of an emergency another person's presence may be essential, always keep one hand in your pocket when anywhere around a powered line-connected or high voltage system, wear rubber bottom shoes or sneakers when working on LCD television, work area should be set-up away from possible grounds that you may accidentally contact, don't wear any jewellery or other articles that could accidentally contact circuitry and conduct current, or get caught in moving parts, use an isolation transformer if there is any chance of contacting line connected circuits, if you must probe live, put electrical tape over all but the last 1/16" of the test probes to avoid the possibility of an accidental short which could cause damage to various components, perform as many tests as possible with power off and the equipment un-plugged, avoid working when you are really tired, unplug the TV set from socket outlet after use to avoid burning of fuse or other electronic components as a result of lightening, and if the television clicks with power-on but no other action, check other fuses in the circuit and replace them accordingly. This result is in consonance with [9] . Goldwasser [9] , identified some of the precautionary measures as wearing rubber shoes or sneakers, not wearing jewellery, among others. This is an indication that accident may occur especially electrocution when working in any current device or appliance without wearing protective shoes.

The result of the analysis of the hypothesis revealed that there was no significant difference in the responses of electronic technology teachers and electronic technicians on 29 important precautionary measures that should be taken in maintenance of LCD television faults. However, significant difference existed in the respondents responses on only three items as can be seen from Table 2. The hypothesis of no significant difference was upheld for 29 items, but was rejected for 3 items (see Table 2). Majority of the electronic technology teachers and electronic technicians rated the important precautionary measures that should be taken in maintenance of LCD television faults the same way. The result confirms the findings of Goldwasser [9] , who identified some of the precautionary measures as wearing rubber shoes or sneakers as well as not wearing jewellery.

9. CONCLUSION

Based on the findings of the study, the following conclusions were made: LCD televisions are new in the society and their maintenance should be taught as a separate topic in electronic technology. Electronic technicians should carry out maintenance of LCD televisions by observing the precautionary measures to be taken in maintenance of LCD televisions. This will avoid damaging other components or sections that are working properly.

REFERENCES

1. Bayle, T. (2007) Preventive Maintenance Strategy for Data Centres. American Power Conversion, America.

2. Operations & Maintenance Best Practices (2011) A Guide to Achieving Operational Efficiency Chapter 5, Types of Maintenance Programmes, Release 3.0.

3. Leading Learning Council and Skill, Apprenticeships: 2005 (2011) www.apprenticeships.org.uk

4. Okoro, O.M. (1999) Principles and Methods in Vocational and Technical Education. University Trust Publishers, Nsukka.

5. Federal Republic of Nigeria (2004) National Policy on Education (Revised Edition). Nigeria Education Research and Development Council (NERDC) Press, Lagos.

6. Theraja, B.L. (2007) Basic Electronic Solid States. S. Chand and Company Ltd., India.

7. Theraja, B.L. and Theraja, A.K. (2007) A Text Book of Electrical Technology. S. Chand and Company Ltd., India.

8. Wistrom, E. (2011) Common Vizio LCD TV Problems & Solutions. http://www.vizio.com/contacts/

9. Goldwasser, S.M. (2011) Notes on the Troubleshooting and Repair of Television Sets, Version 3.30.

10. Sony, LCD Digital Colour TV (2010) http://www.sony.net/

11. Weick, K.E. (1991) Organizational Culture as a Source of High Reliability. In: James, R., Achieving a Safe Culture: Theory and Practice. http://www.shef.ac.uk/~mcn/6810/reading/reason-1998-safety-culture.pdf

12. James, R. (1998) Achieving a Safe Culture: Theory and Practice. http://www.shef.ac.uk/~mcn/6810/reading/reason-1998-safety-culture.pdf.

13. Hudson, P. (1999) Safety Culture—Theory and Practice. RTO HFM Workshop on "The Human Factor in System Reliability—Is Human Performance Predictable?", Siena, 1-2 December 1999, Published in RTO MP-032.

14. Ezeji, S.C.O.A. (2004) Basic Principles of Research in Education. Cheston Agency Ltd., Enugu.

15. Anambra State Post Primary School Management Board (2011) Population of Electrical/Electronic Technology Teachers. Statistics Department.

16. Anambra State Ministry of Commerce and Industry (2012) Population of Radio and Television Repairers in Anambra State. Anambra State Radio and Television Repairers Association, Anambra.

17. Gall, M.D., Gall, J.P. and Borg, W.R. (2007) Educational Research an Introduction. 8th Edition, Pearson.

18. AU Optronics Corporation (2010) Product-Category Rules (PCR) for Preparing an Environmental Product Declaration (EPD) for TFT-LCD Televisions. AU Optronics Corporation, Taiwan.

19. Lewis, K. (2004) Sony LCD Television Service Manual. Sony Corporation Sony UK Service Promotions Department, UK.

CHAPTER 2

Liquid Crystal Microlenses for Autostereoscopic Displays

José Francisco Algorri *, Virginia Urruchi, Braulio García-Cámara and José M. Sánchez-Pena

Electronic Technology Department, Carlos III University of Madrid, Avenida de la Universidad 30, Leganés E28911, Spain

ABSTRACT

Three-dimensional vision has acquired great importance in the audiovisual industry in the past ten years. Despite this, the first generation of autostereoscopic displays failed to generate enough consumer excitement. Some reasons are little 3D content and performance issues. For this reason, an exponential increase in three-dimensional vision research has occurred in the last few years. In this review, a study of the historical impact of the most important technologies has been performed. This study is carried out in terms of research manuscripts per year. The results reveal that research on spatial multiplexing technique is increasing considerably and today is the most studied. For this reason, the state of the art of this technique is presented. The use of microlenses seems to be the most successful method to obtain autostereoscopic vision. When they are fabricated with liquid crystal materials, extended capabilities are produced. Among the numerous techniques for manufacturing liquid crystal microlenses, this review covers the most viable designs for its use in autostereoscopic displays. For this reason, some of the most important topologies and their relation with autostereoscopic displays are presented. Finally, the challenges in some recent applications, such as portable devices, and the future of three-dimensional displays based on liquid crystal microlenses are outlined.

KEYWORDS

autostereoscopic display; spatial multiplexing; liquid crystal microlenses

1. INTRODUCTION

More than 170 years ago, Charles Wheatstone demonstrated the first stereoscope to the Royal Society (1838) [1]. He illustrated the basic principle of stereoscopic vision, the brain's capacity to achieve depth perception from different images. He combined mirrors to present separate images representing different perspectives of the same object for each eye (Figure 1). A few years later, Rollman proposed the use of anaglyphs in order to display stereoscopic pairs [2]. The barrier technique was later proposed by both Jacobson and Berthier around 1896 and it was first applied by Ives in 1903 [3]. He called it the "Parallax Stereogram". Later, in 1908, a new technique based on using spherical lenses instead of opaque barrier lines were proposed by Lippmann [4]. This was called "La Photographie Integral". Lippman was a man before his time and this invention could not be further developed because of the technical limitations at the time. Some researchers, such as Herbert Ives, simplified this technique in 1930 by using a lenticular lens array [5]. These images were of poor quality due to the production methods available at the time. In the 1950s, three-dimensional (3D) vision was attempted to be displayed in some cinemas by using anaglyphs. The low quality of the content led to a decrease in interest. In the 1970s, the system was improved thanks to the patented system of Stephen Gibson "Deep Vision" [6]; red-cyan anaglyphs could reproduce the color of the skin better than the red-blue or red-green anaglyphs. Despite this, it was not until the 1980s when better results were achieved thanks to IMAX (Image Maximum) high-resolution displays and the use of linearly polarized glasses. The introduction of these systems produced a great commercial impact that increased the interest of research groups in this field.

Currently, 3D technology can be classified into two major groups: one requiring the help of external means, normally glasses (stereoscopic), and another that allows the generation of 3D vision without their need (autostereoscopic). As can be deduced, the use of autostereoscopic devices is the current objective of research groups. These systems were investigated for the first time by Ives (1903) and Lippman (1908), using a barrier system and microlens, respectively. However, technological limitations stopped their development. More recently, this technology has improved considerably, primarily due to new manufacturing processes, such as advanced photolithography techniques, and the reduction of production costs. Furthermore, this technology can now display 3D and different styles of animated visual effects without the requirement of special viewing glasses. Among the multiple techniques on autostereoscopic displays, there are three major lines of research:

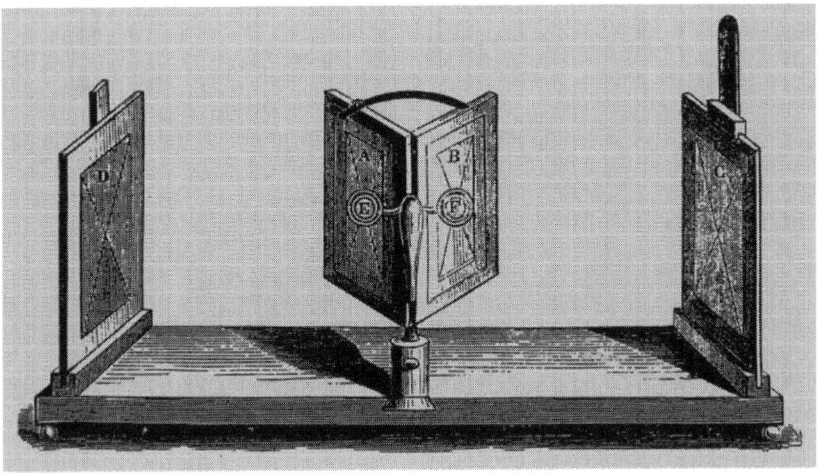

Figure 1. Charles Wheatstone-mirror stereoscope (XIX century) [1]. Public Domain.

Volumetric technique creates the 3D image by projecting the rays onto a volume (Figure 3) or using discrete locations of luminescence within a volume [9]. There is no officially accepted topology because of the variety of volumetric displays. They can be based on the eye persistence (swept-volume display) or using laser light to encourage visible radiation in a gas (static volume). The main problems are the use of movable parts and the low resolution. Moreover, a large amount of data is necessary to store the 3D image.

Electro-holographic is a diffraction-based coherent imaging technique [7]. These systems are based on the phase difference produced when different optical wavelengths are reflected by a certain object. In this type of system, a specific object can be reproduced from a flat 2D display that possess complex transparency representing amplitude and phase values (Figure 2). This technique is one of the most promising because it can reproduce 3D vision with full parallax and without the problems of convergence and accommodation. This is also the most challenging technique because it requires coherent sources and very high resolution spatial light modulators. Interest in this technology has increased exponentially in the last years.

Spatial multiplexing is based on multiplexed 2D images. There are two main techniques: time and spatial multiplexing. In the first case, the 2D image reach one eye at different times, exploiting the persistence of the human visual system to give 3D perception. This technique usually needs spatial light modulators and active shutters [10] or special backlights [11] or simply dynamic displays [12]. The main advantage is the full resolution of the displayed images, while some common problems are moiré patterns [13], crosstalk [14] and flicker [15]. For this reason, this technique is in the research stage and no commercial displays have been produced. Spatial multiplexing techniques apply optical phenomena including diffraction, reflection, refraction and occlusion in order to deviate the images to certain positions in front of the display. Among the different technologies to achieve this effect, refractive elements (microlenses) and occlusive elements (parallax barrier) are the most successful. A parallax barrier consists of masks that contain vertical apertures to cover the light at certain angles (see Figure 4, left). The barrier can be considered a mature technology because commercial devices have already been produced. For example, dynamic barriers based on Liquid Crystal Displays (LCDs) are used to generate black and white columns. Despite its maturity, this technology has considerable drawbacks such as lower brightness, small viewing angles, and crosstalk, which is caused by diffraction and it is difficult to suppress [16]. To improve the function of this type of system, lenticular technology can be used (see Figure 4, right). In this case, microlenses deviate the light to certain directions corresponding to each different eye.

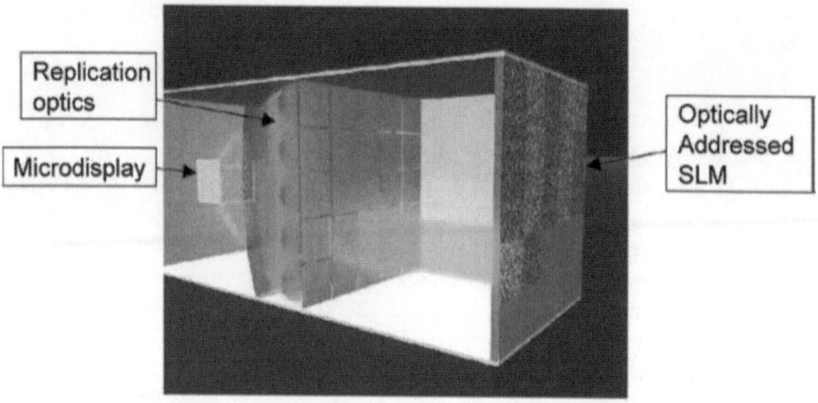

Figure 2. One channel of the active tiling modulator concept of QinetiQ [8]. Reproduced with permission from Slinger, C.; Cameron, C.; Stanley, M. Computer; published by IEEE, 2005.

In this review, a study of the impact of these three technologies over the last few years has been carried out. In Figure 5a, the three main lines of research mentioned above are classified in terms of research manuscripts per year (source: Scopus). Interesting results arise from this survey. Among the three techniques, spatial multiplexing is the one with the largest scientific production. In fact, this technique is the most successful in terms of efficiency and quality. Most of the commercial autostereoscopic displays are based on this technique and some of the most promising research is focused on this line (integral imaging). In Figure 5b, the works devoted to the two main solutions to implement this technology (parallax barriers and microlenses) are studied independently. Due to the disadvantages of parallax barriers mentioned above, the research of this technology has been developed to a lesser extent in comparison to lenticular devices based on microlenses. Holographic technique is still in the research stage. Despite this, as can be seen in Figure 5a, the number of scientific manuscripts in this field has greatly increased in recent years. This is probably the most realistic 3D reproduction but the technical problems are still numerous. The tendency indicates that in two or three years the amount of research will be similar to spatial multiplexing techniques. Finally, volumetric research peaked few years ago, but the survey seems to suggest a decreasing interest in this technique.

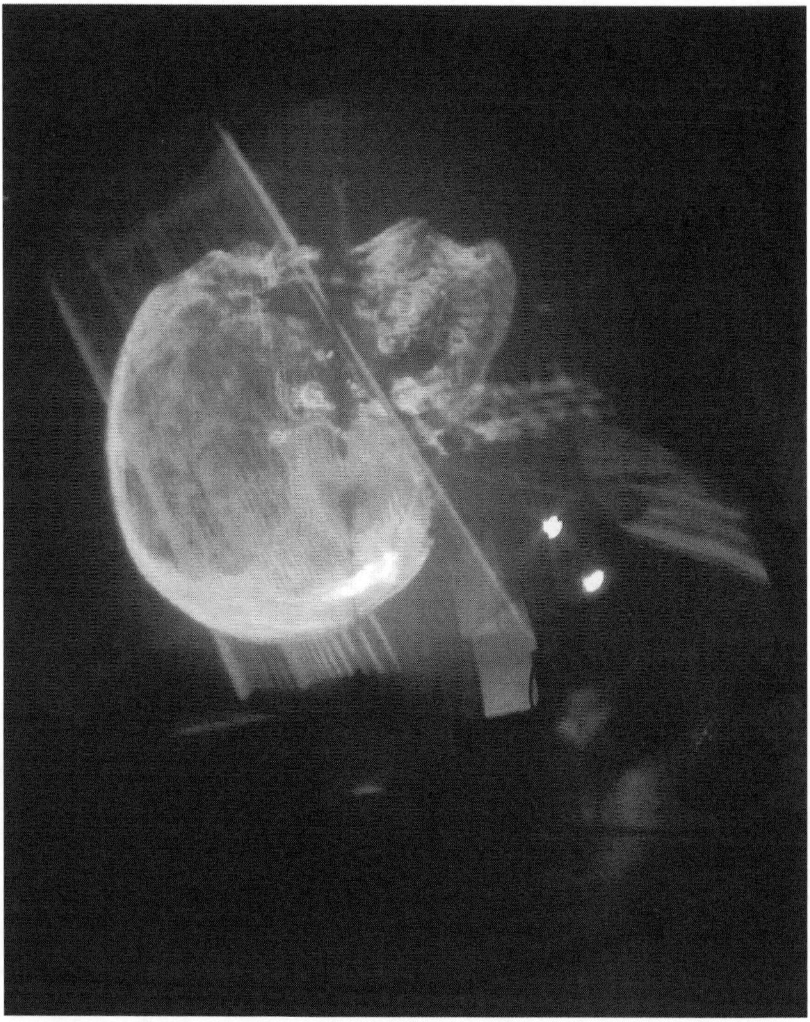

Figure 3. "PerspectaRAD mouse Phantom" by Gregg Favalora. Licensed under CC BY-SA 3.0 via Commons.

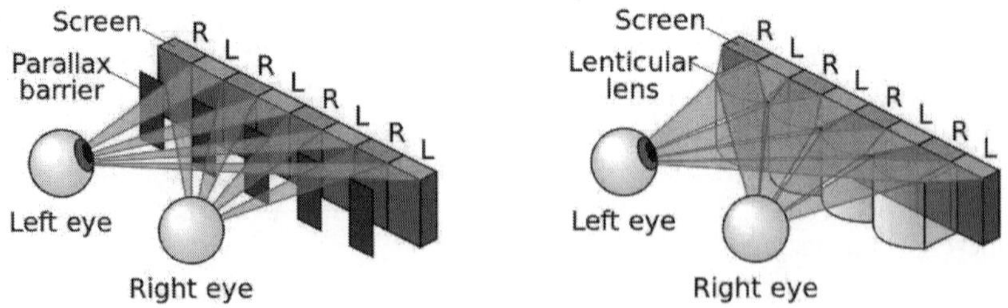

Figure 4. "Parallax barrier vs. lenticular screen" by Cmg Lee. Licensed under CC BY-SA 3.0 via Commons.

As stated above, the most researched technique is actually spatial multiplexing. Moreover, Liquid Crystal (LC) lenses may be used to improve the characteristics of this technique. In this review, first we discuss the technical issues of spatial multiplexing techniques used actually. Some commercial products and their technical characteristics are presented in Section 2. The use of microlenses is clearly stated. For this reason, in Section 3 a complete study of LC microlenses, focused on their use in autostereoscopic devices, is presented. Finally, a critical analysis is included in Section 4.

2. STATE OF THE ART OF SPATIAL MULTIPLEXING TECHNIQUE

This technique is currently used in several commercial products and seems to be the most promising for giving rise to future products in the market. The basic scheme is based on the optimization of two aspects: the capture of a stereoscopic pair of images stereogram for display on a screen and the correct visualization of these images by the user. This configuration is known as a two-view autostereoscopic display. The stereograms consist of images (corresponding to the same scene viewed from two different angles) that are intertwined in columns of pixels. At this point, a system capable of directing the path of light of each column of pixels, to the corresponding eye, is necessary. Then, the brain interprets the signals and generates a 3D virtual image. However, there are some practical problems. The viewer must stay in the right position or a pseudoscopic image is produced. Moreover, eyestrain may occur caused by a mismatch between convergence and accommodation. This is an unnatural condition that disrupts the correlation between vergence and focal distance [17]. The larger the depth is, the larger the visual strain becomes. These limitations can be overpassed by increasing the number of views and using full parallax (multi-view [18] or integral imaging [19]). In multiview displays, multiple different images are presented in front of the display. Still, the number of views is too small for continuous motion parallax. Some proposals have been made to minimize the transitions between views [20]. Still, no full parallax is provided.

Integral imaging has both characteristics, motion and full parallax. Nevertheless, such technique proposed by Lippman in 1908 have been recently possible from a technical point of view. The need of ultra-high resolution displays, fast processors for processing the large amount of data and a poor quality of the microlens arrays are some reasons of this delay. Integral imaging is the most promising of all spatial multiplexing techniques as it achieves to reproduce a 3D field view similar as it is reproduced in nature [19].

The two main approaches to achieve spatial multiplexing: parallax barriers [21] and lenticular systems [22]. The barriers can be considered a mature technology because some commercial devices based on them have been produced. For instance, dynamic parallax barriers were proposed in order to change the properties of the virtual image (3D distance, number of views, etc.) [23]. This technology has considerable drawbacks that can be overpassed by using lenticular lenses.

As mentioned above, parallax barriers and fixed lenticular systems (no based on LC), can already be found in commercial products. This year, (6–9 January 2015), Toshiba presented at the "2015 International Consumer Electronics Show" in Las Vegas a new display using LC lenses. Toshiba plans to fuse the new technology with a partial 2D/3D switching function which can be applied to any screen size and position, aiming for its rapid commercialization in industrial and medical products that require glasses-free high-definition 3D displays. Despite this announcement, the information of the technology behind is still unknown. This is one of the first commercial displays using LC lenses. Table 1 also shows two representative examples of 3D displays already on the market, developed by Sharp and Toshiba.

The device of Sharp uses a dynamic parallax barrier that only a few years ago was in the research stage. It is based on the use of LCs and it has many advantages, such as 2D/3D switching, multiple views, variable focal length, wide viewing zone, etc. The Toshiba display is based on a set of microlenses of high quality, probably manufactured with some kind of resin or polymer. This display uses a specific eight-core processor that allows calculations of the required views as a function of the number of observers. Some common disadvantages of spatial multiplexing systems have been solved in different ways:

- The loss of resolution is solved through very high-resolution displays (QFHD (quad full high definition), 3840 × 2160 pixels). Moreover, the lenses are placed with a small angle to distribute the loss of resolution vertically and horizontally [24].

- Another common problem is the non-uniformity of the light coming from pixels. The problem is caused by the darker spaces of the pixels that generate bright and dark areas. An observer looking at different angles will notice a change in the image brightness under certain positions. There are three methods to solve this: place the lenses with a convenient angle, adapt the focal distance, and modify the width of the lenses to obtain fractionated views [25].

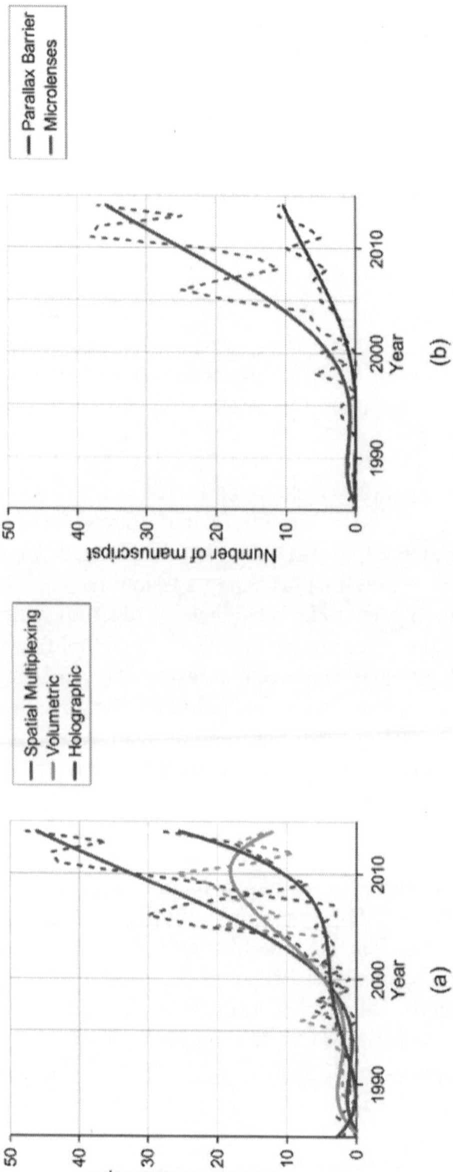

Figure 5. Number of manuscripts as a result of the search in Scopus database (Elsevier). The search was restricted to only the titles of the manuscripts: (**a**) Autostereoscopic technologies and (**b**) Spatial multiplexing technologies. Queries for each technology were "holographic and display and 3D (or three-dimensional or autostereoscopic)", and "volumetric display" and for spatial multiplexing, "parallax barrier display" and "lenticular (or lenses or microlenses) and display". Dashed lines are real data, while solid lines are tendency lines (polynomial fit). The fitting is done by an orthogonal polynomial with four degrees.

- Another significant problem are aberrations caused by manufacturing defects. Aberrations cause a non-uniform distribution of the intensity. However, thanks to improvements in various processes, the non-uniformity of the lenses can reduce the crosstalk to values ranging from 2%–7% [26].

Table 1. Commercial examples of autostereoscopic devices based on spatial multiplexing.

Product	Company	Technology	Resolution	Mode	Year
	Sharp 3.4″ (3DS)	Dynamic parallax	480 × 854	2D/3D	2011
	Toshiba 55″ (55ZL2G)	Fixed Lenticular array	3840 × 2160 (4K)	2D/3D	2012
Unknown	Toshiba 15″ (Unknown)	LC Lenticular array	4K	2D/3D/4K	2015–2016

Previous systems are patented or commercially available. However, there are still many problems to be solved. Thus, autostereoscopic vision has become a topical issue that requires numerous research resources. In the field of lenticular arrays, the possibility of using LC lenses is being actively investigated. Within LC, lenses can be separated into active (birefringent active) and passive (activated by polarization). Table 2 summarizes a comparison between technologies. Brightness 2D and 3D indicate the total brightness of the display when 2D or 3D images are reproduced, respectively. In autostereoscopic vision, the crosstalk parameter is used to measure the quality of the generated stereoscopic image [27]. Voltage 2D and 3D are the necessary applied voltage to maintain a 2D and a 3D image, respectively. Collected data are the best characteristics found in the bibliography for each of them. It is likely that one of these topologies is part of the new Toshiba display. It must be borne in mind that fixed microlens technologies and fixed barriers can be found implemented in commercial products, while devices based on LC are still part of the research field. Given the great commercial impact of these technologies, they will probably be marketed soon.

In this study, different existing technologies for autostereoscopic applications have been explored. It has been found that the only system that allows a tuning of the 3D distance is the method of dynamic parallax barrier with the disadvantages that this system entails. Devices that employ LC lenses are only designed for 2D/3D operation and usually are based on curved surface (explained in next section). There are numerous techniques for manufacturing LC lenses that could be of clear advantage to use in autostereoscopic displays. For instance, patterned electrode is the most promising technique for small pixel sizes. The main problem caused by the low speed of LC lenses could be solved by using stacks of lenses [28]. Aberrations that are a common drawback in these devices could be compensated for by using an aberration compensator devices for rectangular apertures [29]. Besides, the temperature dependence of the LC birefringence can be controlled by special temperature sensors for this type of structures [30,31]. In some cases, when the pixel size is bigger than certain diameter (>200 μm) other topologies would be necessary (curved surface or modal control) [32].

Table 2. Comparison between different autostereoscopic technologies.

Characteristic	Fixed Parallax	Dynamic Parallax	Passive Parallax	Fixed Lenses	Active Lenses	Passive Lenses
Brightness 2D *	45%	45%–85%	45%	>95%	>95%	>95%
Brightness 3D *	45%	<20%	<20%	>95%	>95%	>95%
Contrast	1:1000	-	-	1:1500	-	-
Crosstalk	>2%–3%	>2%–3%	>2%–3%	<1%	<1%	<1%
Thickness	0.5 mm	1.13 mm	1.65 mm	0.7 mm	1 mm	1.1 mm
Voltage 2D	-	0 V	0 V	-	Depend on the topology	0 V
Voltage 3D	-	3.3 V	3.3 V	-	0V	3.3 V
Switching time	-	<100 ms	<100 ms	-	Several seconds	<100 ms
Multiplexing of 2D/3D areas	-	Differences ×3–×5 in brightness	Differences ×2 in brightness	-	Yes	Yes

* Over a maximum brightness of 500 cd/m^2.

3. LIQUID CRYSTAL MICROLENSES FOR AUTOSTEREOSCOPIC DISPLAYS

A homogeneous conventional lens has two physical characteristics that contribute to the way in which such lenses modify a wavefront passing through it: the difference between the refractive index of the lens material and the surrounding environment, and the curvature of their interfaces. In addition, it is known that when light travels through a non-homogeneous medium, the wavefronts decrease their speed in the optically dense regions and accelerate in areas of lower density. Therefore, it should be possible to design a lens without curvature but with a material that has a gradient in the refractive index; this device is known as a GRIN (Gradient Index) lens. This is the main operating principle of LC lenses. Liquid crystals are optical anisotropic materials with birefringence that can be controlled by an applied voltage. Thus, most of the proposed LC lenses have a graded refractive index produced by a gradual voltage across the lens diameter. These kinds of lenses have potentially the same applications as conventional fixed lenses, but also have some interesting properties for their practical use, for example, small size, lightweight, low driving voltages, low power consumption and transmissive/reflective operation modes.

At the end of the 1970s, some pioneer research, which gave rise to the first proposals of adaptive lenses based on LC materials, were carried out. Berreman et al. with patent application in 1977 [33] and Sato et al. in 1979 [34] (Figure 6) were the pioneers. The first adaptive lens was formed by a cavity of glass with a certain curvature filled with LC. This device had a low response time due to the LC layer thickness. In addition, another problem was the molecular orientation within this thick and curved cavity. This is the reason why this particular technique was not further developed.

Two years later, in 1981 [35], a group from Syracuse University made a LC cylindrical lens with electronic control working as an adaptive optical element. This type of lens employed a large number of electrodes to achieve a proper voltage gradient. This new concept of lens gave rise to other research works [36] and subsequently to spherical lenses [37].

At the end of the 1980s, new research works demonstrated LC lenses with micrometric diameters for the first time. First references are from Sato et al. [38,39]. In this time, the first Fresnel lenses were also proposed [40]. This configuration reduced the required thickness and increased the achievable diameters.

Even though LC lenses were first reported more than 30 years ago, they remain an active field of research today. In general, the applications of LC lenses, regardless of whether they are microlenses or not, are numerous, as are the applications of fixed lenses or GRIN lenses, with the great advantages in weight and volume reductions and a focal length tunable by voltage. For instance, LC lenses have been proposed to work in imaging systems of portable devices. In this field, LC lenses can help realize auto-focusing systems and optical zoom systems for portable devices, such as cell phones and cameras [41]. In addition, it has been proposed for pico-projection systems, helping to electrically adjust the focusing properties of the projected image without a mechanically adjusting position of a projection lens [42]. In the case of holographic projection systems, a LC lens can also help to correct the mismatch of the chromatic image size which is very important for a full-color holographic projection system [43]. Liquid crystal lenses can also be used as concentrators and a sun trackers in concentrating photovoltaic (CPV) systems [44].

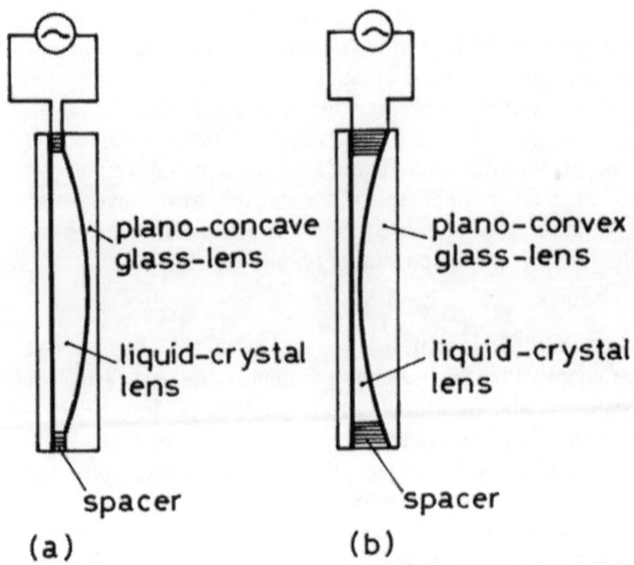

Figure 6. First proposals of LC lenses: (**a**) Plano-concave lens and (**b**) Plano-convex lens [34]. The Japan Society of Applied Physics (JSAP). Reprinted with permission from Sato, S. Japanese Journal of Applied Physics; published by IOP Publishing, 1979.

It is worth mentioning the field of bio-optics, for example, the use of lenses for medical instrumental in applications like endoscopy. A LC lens can be adopted to electrically enlarge the depth-of-field of the endoscopic system [45]. Moreover, the lens power of LC lenses is not only electrically tunable, but it can also be positive or negative, which has interest for ophthalmic lenses. In this case, lenses with millimeter diameter are required [46]. Liquid crystal lenses can correct myopia impact. Commercial glasses based on LC lenses were developed in 2011. They have a switchable focal distance [47] solving the problem for patients that require different types of glasses for different activities. Unfortunately, the company went bankrupt this year (2015) due to a high rate of return (battery issues and some defective devices). Liquid crystal lenses can also be a kind of "extra-artificial crystalline lens" to compensate the degradation of the crystalline lens of aging eyes or eye accommodation [48]. Some important advances in this field can be produced by LC lenses working with unpolarized light [49], which produces a considerable increase in the optical efficiency.

In short, in recent years, there have been lots of developments tailored to the needs of different applications. Over this period, many new topologies have been proposed, such as polymer gel stabilization, patterned electrode, curved surface, Fresnel lens or modal control. At micrometric scale, and for use in autostereoscopic applications, the following should be highlighted: curved surface, patterned electrode and modal control. The next three sections study these topologies in detail. Finally, the last section summarizes the different fabrication methods and characteristics of these topologies.

3.1. Curved Surface

This category includes all lenses having at least a non-planar surface. Most of the autostereoscopic prototypes based on LC microlenses use this topology. The criteria of classification of lenses from the following list is based on the type of structure and material used in the fabrication process:

- Curved ITO: Consists of a curved electrode of indium titanium oxide (ITO). This structure has the same problems as dual-voltage lenses which require a complex fabrication process [50]. Moreover, no reports at micrometric scale have been reported. For these reasons, they are less relevant to this list.

- Curved glass: These lenses were the first LC lenses, a cavity of glass [51]. Although it could be guessed that this old design would be overpassed by new technologies, it is, surprisingly, still used. The main reason is the application of this design to autostereoscopic devices. As mentioned above, this particular technique was not further developed because it had low response time and problems of lack of homogeneity. However, new proposals have reported LC to work as microlenses (immersed polymer microlenses) [52]. This reduces the necessary thickness. Moreover, techniques of LC multilayers further reduce the thickness and the inhomogeneity (Figure 7). When a dielectric layer and a LC layer are sandwiched between two continuous electrodes, the thickness variance of the dielectric layer gives rise to an inhomogeneous electric field in the LC layer [53].

- Curved Photoresin: In this approach, the photoresin has a spherical or cylindrical shape and is surrounded by the LC material [54]. Its main disadvantage is the complex fabrication process. Another option is to produce spherical cavities of photoresin and filled them with LC. This topology has been proposed to work in autostereoscopic devices with switchable 2D/3D mode:

    - Active birefringent lenses: A typical structure is shown in Figure 8 [55]. The cavity formed by the photoresin is filled with LC.

Unswitched, a polarized light is affected by the extraordinary refractive index, greater than that of the photoresin (a positive lens is formed). When the LC is switched by an external electrical field, the polarized light is affected by the ordinary refractive index, similar to that of the photoresin (the light passes through without deviation). The main disadvantage is the high operating voltage (50–100 V).

- Polarization Activated Microlenses: In this structure, two LC layers are used (Figure 9) [22]. One layer acts as the lens (nematic LC) and the other control the polarization by a twisted nematic (TN) LC. The TN LC cell change the polarization of the incoming light so it is affected by the extraordinary or ordinary refractive index of the nematic LC layer. This reduces considerably the operating voltage (only the TN cell is switched) but complicates the fabrication process. This type of lens only switches between focusing and non-focusing states, rather than tuning in a continuous focusing range like other LC lenses. For this reason, this structure has been proposed to work in a switchable 2D/3D mobile phone display [56]. Another option is to create polymeric lenses and use a similar structure [57]. The advantages are the switching speed and a reduction in crosstalk.

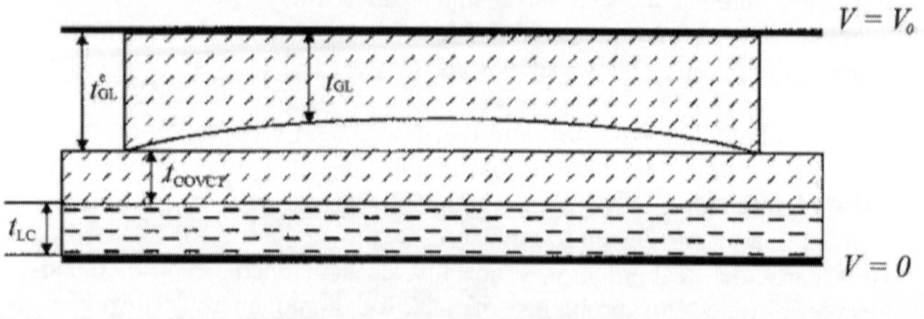

Figure 7. Multilayer liquid crystal lens [53]. Reprinted with permission from Wang, B.; Ye, M.; Sato, S. Applied Optics; published by OSA Publishing, 2004.

In summary, curved surface topology has the problems of high operating voltage or complex fabrication process. Liquid crystal alignment becomes a challenge near the curved surface. Conventional techniques like mechanical rubbing and photo-alignment do not provide satisfactory alignment quality on the curved surface. To overcome these problems, patterned electrode technique could be used. It makes use of the fringe fields produced through a dielectric layer to produce a phase profile within a LC lens. The optical properties depend on the relation between the lens aperture and the LC thickness. This configuration requires lenses with diameters smaller than 200 μm (it depends on the LC layer thickness). This technology is explained in the next section.

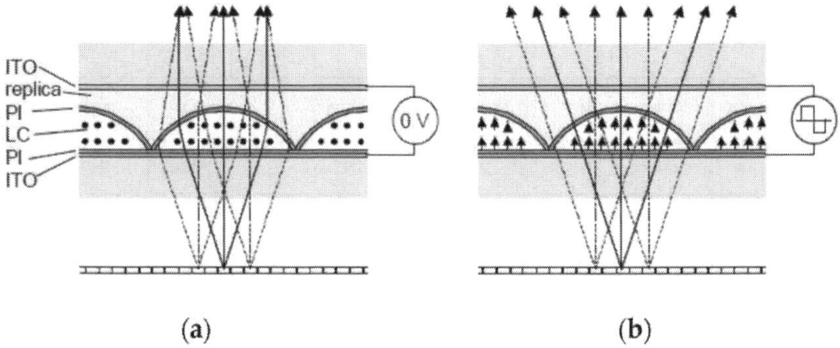

Figure 8. Active birefringent lens: (**a**) without voltage and (**b**) with voltage [55]. Reprinted with permission from Willemsen, O.H.; de Zwart, S.T.; IJzerman, W.L.; Hiddink, M.G.H.; Dekker, T. International Society for Optics and Photonics; published by SPIE, 2006.

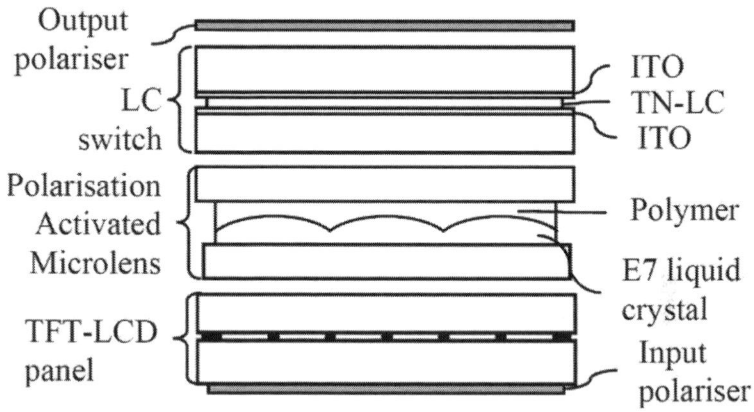

Figure 9. Polarization activated microlenses [22]. Reprinted with permission from De Boer, D.K.G.; Hiddink, M.G.H.; Sluijter, M.; Willemsen, O.H.; de Zwart, S.T. International Society for Optics and Photonics; published by SPIE, 2007.

3.2. Patterned Electrode

As mentioned above, patterned electrode technique was the first proposed method to obtain LC lenses. Despite this, this technique has reinvented itself and is still present in many novel proposals. At the beginning, only a series of electrodes in contact with the LC were used to create a voltage gradient [58]. The first microlenses were fabricated by Nose and Sato in 1989 using hole patterned technique [38]. This technique is made up by several patterned circular holes in the upper electrode (Figure 10). If the diameter is small enough, the fringe fields are capable of generating a lens-like phase profile in the LC layer. Years later, other models were proposed based on the inclusion of a dielectric between the electrode and the LC layer. Thanks to this configuration, the fringing field near the electrode edges (that causes unwanted LC twist alignment) is avoided. This technique, proposed in 2004, was known as dual voltage [59]. The main disadvantage was the high operating voltage (>50 V) caused by the space between the electrode and the LC. When a high optical power is required, multilayered structures could be used [48].

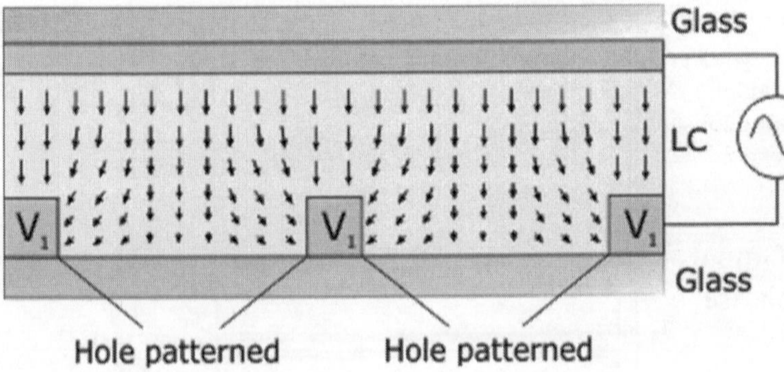

Figure 10. Structure of a liquid crystal microlens based on hole patterned technique.

Recently, a lot of research based on this technology has been presented. Most of them place the electrodes next to the LC layer as the first microlens (Figure 10). For example, tetragonally hole-patterned electrodes [60]. In that work, each LC cell in the lens array behaves like a cylindrical or spherical lens by electrically adjusting the applied voltages. Another recent proposal is a LC lens with two divided and double circularly hole-patterned electrodes [61]. The radially and axial symmetrically distributed refractive index of the hole-patterned aperture can be obtained by controlling the two divided circularly hole-patterned electrodes and the double circularly hole-patterned electrodes.

The requirement of working with polarized light is an issue that reduces the final brightness in autostereoscopic displays based on LC microlenses. Polarization independent adaptive microlenses have also been proposed. These proposals avoid the use of polarizers increasing the optical efficiency. There are different approaches, including blue-phase LC [62] and nematic LC [63]. Circular hole-patterned electrodes with tunable coaxial bifocals (CB) synthesized via photopolymerization of LC cells have also been demonstrated recently [64]. Despite the photo-polymerization method, the bi-axial confocal LC lens can also be fabricated by hole-patterned structure [65]. Devices manufactured with this method consist of two types of tunable CB LC lenses fabricated via different photocurable processes. Another recent proposal has been a hole patterned electrode structure with ultrathin glass slabs [66].

In the field of autostereoscopic devices, some interesting proposals have been made recently. For example, a scanning Multi-electrode Driven Liquid Crystal (MeD-LC) lens, achieved to reproduce several views in a 3D display [67]. This configuration is based on a parallel electrode pattern. By applying a sequentially voltage, a lens-like shape is formed under the electrodes with voltage. Thus, the lens could move (shift) sequentially on the horizontal direction to project the images to different viewing angle. Another work propose the use of patterned microlenses with sizes similar to the pixel pitch to obtain a full resolution autostereoscopic display [68]. Another original contribution is the proposal of this type of lenses to work in autostereoscopic devices, demonstrating the focal length tunability, the deviation of images [69] and the tuning of the 3D distance. Some recent devices have the ability to display vertical and horizontal views; to achieve this function, a hole patterned electrode on each substrate has been manufactured. The substrates have been arranged with the comb type electrodes in orthogonal position, besides complex electrical signals are used [70]. The result is a cylindrical LC microlens array with rotary optical power and tunable focal length capability. A focal distance tunability ranging from 0.2 to 1.2 mm in a 125 μm lens is obtained. This would produce a 3D distance from 20 to 80 cm, perfect for a mobile application. It is important to note that this new configuration also leads to minimum aberrations. In the case that some aberrations appear, a tunable compensator device based on hole patterned electrode has been demonstrated in [29]. It is demonstrated how some classical aberrations as spherical or coma are controlled with voltage more than –1 and 2 waves, respectively. Finally, other research works have been focused on modeling the electric field distribution in order to have optimal designs [32,71]. These studies reveal a critical interrelation between the structural parameters of the microlens in order to distribute the voltage with a parabolic profile. This study concludes that depending on the LC birefringence, a ratio of 2.4 between thickness and diameter of a microlens is stablished for MDA 98-1602 nematic LC from Merck is necessary to manufacture microlenses. For example, in case of multiview displays for diameters of 100 μm, the necessary thickness would be 42 μm. Large thicknesses could limit the switching speed of the final device.

Due to the constant reduction of the pixel size, it seems evident that this topology is going to be protagonist in future products. Despite this, as mentioned above, there are certain cases where the necessary thickness is too thick. One technique that could be used to correctly distribute the voltage in this specific case is modal control.

3.3. Modal Control

Modal control technique has become an alternative for lens design because overpasses the main drawbacks of previous techniques. A key advantage is the driving method, with only one voltage control at very low values (<3 V).

Moreover, when the established ratio between the microlens diameter and thickness is not fulfilled, using this technique instead of patterned electrode configuration is recommended (the fringe fields do not reach the microlens center). First research on modal control was reported by Naumov et al. [72]. This technique consists of generating a radial graded refractive index across the lens aperture by using a layer of high sheet resistance (MΩ/sq), deposited onto the pattern electrode, as a control electrode. Sheet resistance of the control electrode is a key design parameter. Its value must be in the range 100 kΩ/sq to few MΩ/sq for lens diameters on the order of millimeters [73]. This layer creates a voltage divider with the LC impedance causing a hyperbolic voltage gradient across the lens aperture [17]. Different materials have been reported to act as a control electrode layer, Poli(3,4-etilenodioxitiofeno) (PEDOT) [74], thin ITO layers [75] (10 MΩ/sq [76]) and titanium oxide films [77] are some of them. A diagram of a modal liquid crystal cylindrical lens is depicted in Figure 11.

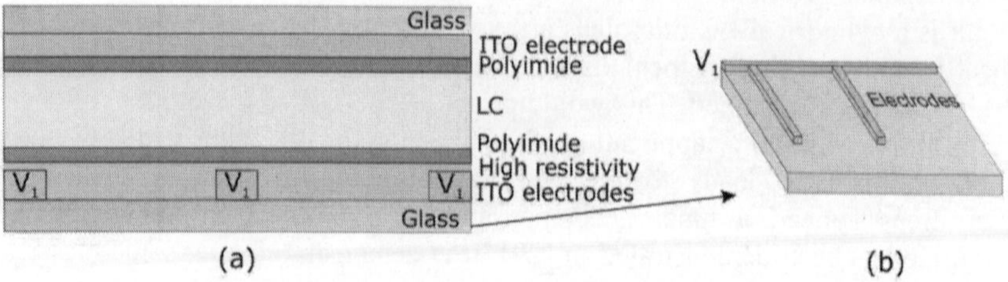

Figure 11. Structure of a liquid crystal cylindrical lens based on modal control technique: (**a**) 2D view and (**b**) 3D view of the bottom substrate, detail of the electrode pattern.

Indeed, it is not a simple task to obtain exact thicknesses of the high impedance layer; setting a thickness value must offer a good compromise between impedance (which affects lens refractive index gradient) and transparency (which influences lens quality). On the other hand, control voltages must have optimized the amplitude, frequency and shape. Due to the capacitive effect of the LC, LC layer impedance is frequency dependent. In addition, one of the main drawbacks of systems using these optical arrangements is the generation of aberrations. Thus, the suitable choice of the voltage shape (set of harmonics) can minimize phase aberrations caused by both, the non-ideal distribution of the electric field and the nonlinear electro-optic response of the LC with voltage [77]. For this reason, some systems that control the frequency, amplitude and duty cycle have been proposed to control the phase profile [78]. Novel structures, based on symmetric electrodes, have also been proposed to reduce aberrations [74]. In this case, spherical aberration range from −0.4 to 0.4 waves and comma range from −0.2 to 0 waves. One application worth mentioning is a wavefront modulator based on modal control technique [79]. In addition, a modal LC lens is commercially available at Flexible Optical B.V. [80].

Unfortunately, the fabrication of modal lenses at micrometric scale implies a technological challenge. The main problems are difficulties in finding layers of 0.1–1 GΩ/sq and the fact that, at some point, the fringe fields distributed by LC are sufficient to create a voltage distribution without the need of a high-resistivity layer.

For autostereoscopic applications, there have been several proposals. In [81,82], a system that enables rotation of the 3D image is demonstrated using this approach (Hi-R LC). Despite this, the resistivity is not high enough to create a voltage gradient when all electrodes are at the same voltage. They need a ground voltage at the center of each lens to have a voltage gradient, reducing the advantage of tunability of modal LC lenses. Other works, propose the use of polymer stabilized blue phase LC materials to obtain fast-response LC lens for 3D displays [83]. Another advantage of these types of systems is that they are polarization independent.

Recently, a cylindrical microlens array with a new manufacturing protocol based on a set of two layers (a dielectric layer of SiO_2 and a metallic layer of nickel) for lenses with modal driving was reported [27]. This array consists of a set of lenses, each 570 μm in diameter whose behavior is based on a nanometric layer of nickel. This layer determines the voltage distribution across the lens diameter; without it, each lens would work as a hole-patterned lens and the voltage would drop at the lens edges. Modal lenses exhibit sometimes some spherical aberration because of the high resistivity of the manufactured layer. However, this aberration can be reduced by increasing the thickness of the metallic layer [84]. The lenses of this contribution were demonstrated to work in an autostereoscopic system and validated through 3D contrast measurements. A 3D distance of 30 cm is obtained for $7.5V_{RMS}$. Unlike other proposed systems [22,51], modal LC lenses have a wide range of focal length tunability, so these systems could adjust the 3D distance over a wide range.

As mentioned above, the necessary sheet resistance is too high in modal microlenses. In order to achieve this value using Nickel, the percolation limit has to be reached. Another option is the use of amorphous Silicon (instead of SiO_2). As it is reported in [85], when a layer of amorphous Silicon is sputtered with Nickel and then is crystallized (with high temperature), the sheet resistance reaches GΩ/sq [86].

3.4. Fabrication Methods of Liquid Crystal Microlenses for Autostereoscopic Applications

In this section, a comparative study of various methods of microlenses fabrication and some important characteristics, described in Section 3.1, Section 3.2 and Section 3.3, is included in Table 3. This table describes the type, topology, certain keys of the fabrication process, the necessary voltage to have the optical power, the switching speed and the requirement of a polarized light source.

Table 3. Comparison between different LC microlenses characteristics.

Type	Topology	Fabrication Process	Voltage 3D	Switching Speed	Polarized Light
Curved Surface	Curved ITO	Sputtered ITO	40–140 V_{RMS}	~1 s	Yes
	Curved Glass	Immersed polymer microlenses	10–20 V_{RMS}	~2 s	Yes
	Curved Photoresin Active	Complex fabrication process, several layers and TN LC in active cells	Depend on the topology	Several seconds	Yes
	Curved Photoresin Passive		3.3 V_{RMS}	<100 ms	Yes
Patterned Electrode	Hole patterned	Simple holes patterned	2–3 V_{RMS}	Several seconds	Yes
	Bi-axial confocal	Special alignment layers	3–10 V_{RMS}	Unknown	No
	Blue-Phase	The blue-phase has to be synthetized	50–100 V_{RMS}	Several μs	No
	MeD-LC	Multiple electrodes	5–10 V_{RMS}	<100 ms	Yes
	Rotary cylindrical	Orthogonal electrodes	2–3 V_{RMS}	Several seconds	Yes
Modal control	Hi-R LC	Difficult to obtain the high resistivity layer	~2 V_{RMS}	~0.6 s	Yes
	Cylindrical microlenses		~7.5 V_{RMS}	Several seconds	Yes

4. CONCLUDING REMARKS AND THE FUTURE OF AUTOSTEREOSCOPIC DISPLAYS

Thanks to the commercial success of some stereoscopic film productions, 3D vision has acquired a growing importance in audiovisual industry in the past ten years. This great commercial impact has produced an exponential increase in 3D vision research. The technological transfer to commercial applications is bigger than in any other field. Despite this, the first generation of autostereoscopic displays failed to generate enough consumer excitement. The main reasons were little 3D content and performance issues. In this sense, the use of LC microlenses brings some important improvements. There are numerous techniques for manufacture LC lenses that could be of clear advantage to use in autostereoscopic displays. For this reason, some topologies of LC microlenses that were in the research stage only a few years ago, will be on the market for large displays soon. For example, curved surfaces seem to be used in a future release of a 15″ autostereoscopic display. Patterned electrode is the most promising for small pixel sizes, while modal control could be used for multiview displays where the required microlenses have diameters bigger than the previous technique.

In the last year, the use of portable devices has grown rapidly. Everything indicates that in the near future most of the multimedia content will be displayed in mobile phones or tablets. For this reason, many experimental and theoretical research groups worldwide have actively worked on alternative solutions for 3D vision in portable devices in which no external devices are required. For these kinds of devices, some aspects have to be taken into account:

- The observer usually has the device in his own hands, so the 3D distance is considerably lower than with big displays. This requires a high optical power.
- Every observer has different physical characteristics, so tunability of this distance is especially required for a useful device.
- Aberrations of the optical elements used in this type of systems are a problem that has to be solved in order to reduce the crosstalk.
- The ability to display vertical, as well as horizontal, views is an added value of autostereoscopic displays.

Liquid crystal lenses are excellent candidates to implement alternative solutions: first, for mitigating some of the identified drawbacks; and second, for creating new gadgets with new capabilities that are not trading currently in the market. The high optical power requirement can be solved using thick samples or multilayered lenses when the response time is critical. The tunability is an inherent characteristic of some configurations, such as patterned electrode. Aberrations can be compensated for with some novel devices also based on LC. The ability to display vertical and horizontal views, while maintaining the tunability characteristic, has been recently solved. In conclusion, the use of spatial multiplexing technique would be possible in portable devices when LC devices are used.

Despite this, the future of autostereoscopic displays seems to be with full parallax and multiview displays. One technology that displays 3D light is integral imaging. Liquid crystal lenses could be of important use in integral imaging. The use of spherical arrays of patterned electrode microlenses with different focal lengths, tunable and controlled aberrations could be the key to achieve displays with high enough quality to reach the market. Still, spatial multiplexing has the bottleneck of final resolution, something that probably could be solved in the future with the introduction of pixels with nanometric size.

ACKNOWLEDGMENTS

This work was supported in part by Ministerio de Economía y Competitividad of Spain (grant No. TEC2013-47342-C2-2-R) and the R&D Program SINFOTON S2013/MIT-2790 of the Comunidad de Madrid.

AUTHOR CONTRIBUTIONS

All authors contributed to scientific discussion and critical revision of the article.

CONFLICTS OF INTEREST

The authors declare no conflict of interest.

REFERENCES

1. Wheatstone, C. Contributions to the Physiology of Vision.—Part the First. On Some Remarkable, and Hitherto Unobserved, Phenomena of Binocular Vision. Philos. Trans. R. Soc. Lond. **1838**, 128, 371–394.

2. Rollmann, W. Notiz zur Stereoskopie. Ann. Phys. **1853**, 165, 350–351.

3. Ives, F.E. Parallax stereogram and process of making same. U.S. Patent 725,567, 14 April 1903.

4. Lippmann, G. Epreuves reversibles donnant la sensation du relief. J. Phys. Appl. **1908**, 7, 821–825.

5. Ives, H.E. Parallax panoramagram made with a large diameter lens. J. Opt. Soc. Amer. **1930**, 20, 332–342.

6. Gibson, S. Stereoscopic apparatus and method of using same. U.S. Patent 4,295,153, 13 October 1981.

7. Yaras, F.; Kang, H.; Onural, L. State of the Art in Holographic Displays: A Survey. J. Disp. Technol. **2010**, 6, 443–454.

8. Slinger, C.; Cameron, C.; Stanley, M. Computer-generated holography as a generic display technology. Computer **2005**, 8, 46–53.

9. Langhans, K.; Bahr, D.; Bezecny, D.; Homann, D. FELIX 3D display: An interactive tool for volumetric imaging. Proc. SPIE **2002**, 4660.

10. Cossairt, O.; Travis, A.R.L.; Moller, C.; Benton, S.A. Novel view sequential display based on DMD technology. Proc. SPIE **2004**, 5291.

11. Chien, K.-W.; Shieh, H.-P.D. Time-multiplexed three-dimensional displays based on directional backlights with fast-switching liquid-crystal displays. Appl. Opt. **2006**, 45, 3106–3110.

12. Liang, H.; An, S.; Wang, J.; Zhou, Y.; Fan, H.; Krebs, P.; Zhou, J. Optimizing Time-Multiplexing Auto-Stereoscopic Displays With a Genetic Algorithm. J. Disp. Technol. **2014**, 10, 695–699.

13. Chen, C.-H.; Yeh, Y.-C.; Shieh, H.-P.D. 3-D Mobile Display Based on Moiré-Free Dual Directional Backlight and Driving Scheme for Image Crosstalk Reduction. J. Disp. Technol. **2008**, 4, 92–96.

14. Shestak, S.; Kim, D. Application of Pi-cells in time-multiplexed stereoscopic and autostereoscopic displays based on LCD panels. Proc. SPIE **2007**, 6490.

15. Travis, A.R. Autostereoscopic 3-D display. Appl. Opt. **1990**, 29, 4341–4342.

16. Mather, J. 3D TV Without glasses. Phys. World **2011**, 24, 33–36.

17. Urey, H.; Chellappan, K.V.; Erden, E.; Surman, P. State of the Art in Stereoscopic and Autostereoscopic Displays. Proc. IEEE **2011**, 99, 540–555.

18. Dodgson, N.A. Optical devices: 3D without the glasses. Nature **2013**, 495, 316–317.

19. Xiao, X.; Javidi, B.; Martinez-Corral, M.; Stern, A. Advances in three-dimensional integral imaging: Sensing, display, and applications [Invited]. Appl. Opt. **2013**, 52, 546–560.

20. Benzie, P.; Watson, J.; Surman, P.; Rakkolainen, I.; Hopf, K.; Urey, H.; Sainov, V.; von Kopylow, C. A Survey of 3DTV Displays: Techniques and Technologies. IEEE Trans. Circuits Syst. Video Technol. **2007**, 17, 1647–1658.

21. Yamamoto, H.; Kouno, M.; Muguruma, S.; Hayasaki, Y.; Nagai, Y.; Shimizu, Y.; Nishida, N. Enlargement of viewing area of stereoscopic full-color LED display by use of a parallax barrier. Appl. Opt. **2002**, 41, 6907–6919.

22. De Boer, D.K.G.; Hiddink, M.G.H.; Sluijter, M.; Willemsen, O.H.; de Zwart, S.T. Switchable lenticular based 2D/3D displays. Proc. SPIE **2007**, 6490.

23. Peterka, T.; Kooima, R.L.; Sandin, D.J.; Johnson, A.; Leigh, J.; DeFanti, T.A. Advances in the Dynallax solid-state dynamic parallax barrier autostereoscopic visualization display system. IEEE Trans. Vis. Comput. Graph. **2008**, 14, 487–499.

24. De Zwart, S.T.; Siebe, T.; Hiddink Martin, G.H. Autostereoscopic image output device. U.S. Patent 20,110,001,803, 6 February 2009.

25. Pezzaniti, J. Autostereoscopic 3-D display. U.S. Patent 20,070,014,023, 3 July 2006.

26. Kooi, F.L.; Toet, A. Visual comfort of binocular and 3D displays. Displays **2004**, 25, 99–108.

27. Algorri, J.F.; Urruchi del Pozo, V.; Sanchez-Pena, J.M.; Oton, J.M. An Autostereoscopic Device for Mobile Applications Based on a Liquid Crystal Microlens Array and an OLED Display. J. Disp. Technol. **2014**, 10, 713–720.

28. Chen, H.-S.; Wang, Y.-J.; Chang, C.-M.; Lin, Y.-H. A Polarizer-Free Liquid Crystal Lens Exploiting an Embedded-Multilayered Structure. IEEE Photonics Technol. Lett. **2015**, 27, 899–902.

29. Algorri, J.F.; Urruchi, V.; Bennis, N.; Sánchez-Pena, J.M.; Otón, J.M. Tunable liquid crystal cylindrical micro-optical array for aberration compensation. Opt. Express **2015**, 23, 13899–13915.

30. Algorri, J.F.; Urruchi, V.; Bennis, N.; Sánchez-Pena, J.M. A novel high-sensitivity, low-power, liquid crystal temperature sensor. Sensors **2014**, 14, 6571–6583.

31. Algorri, J.F.; Urruchi, V.; Bennis, N.; Sánchez-Pena, J.M. Liquid Crystal Temperature Sensor Based on a Micrometric Structure and a Metallic Nanometric Layer. IEEE Electron. Device Lett. **2014**, 35, 666–668.

32. Algorri, J.F.; Urruchi, V.; Bennis, N.; Sánchez-Pena, J.M. Using an Analytical Model to Design Liquid Crystal Microlenses. IEEE Photonics Technol. Lett. **2014**, 26, 793–796.

33. Berreman, D.W. Variable-focus LC-lens system. U.S. Patent 4,190,330, 26 February 1980.

34. Sato, S. Liquid-Crystal lens-cell with variable focal length. Jpn. J. Appl. Phys. **1979**, 18, 1679–1684.

35. Kowel, S.T.; Cleverly, D. A Liquid Crystal Adaptive Lens. In Proceedings of the NASA Conference Held at Langley Research Center, Hampton, VA, USA, 18–19 August 1981; pp. 329–339.

36. Cleverly, D.S. Creation of a Lens by Field Controlled Variation of the Index Of Refraction in a Liquid Crystal. Ph.D. Thesis, Syracuse University, New York, NY, USA, 1982.

37. Riza, N.A.; Dejule, M.C. Three-terminal adaptive nematic liquid-crystal lens device. Opt. Lett. **1994**, 19, 1013–1015.

38. Nose, T.; Sato, S. A liquid crystal microlens obtained with a non-uniform electric field. Liq. Cryst. **1989**, 5, 1425–1433.

39. Nose, T.; Sato, S. Optical properties of a liquid crystal microlens. In Proceedings of the International Conference on Optoelectronic Science and Engineering, Beijing, China, 22–25 August 1990.

40. Williams, G.; Powell, N.; Purvis, A.; Clark, M.G. Electrically controllable liquid crystal fresnel lens. Proc. SPIE **1989**, 1168.

41. Ye, M.; Wang, B.; Uchida, M.; Yanase, S.; Takahashi, S.; Sato, S. Focus tuning by liquid crystal lens in imaging system. Appl. Opt. **2012**, 51, 7630–7635.

42. Lin, Y.-H.; Chen, M.-S. A Pico Projection System With Electrically Tunable Optical Zoom Ratio Adopting Two Liquid Crystal Lenses. J. Disp. Technol. **2012**, 8, 401–404.

43. Lin, H.-C.; Collings, N.; Chen, M.-S.; Lin, Y.-H. A holographic projection system with an electrically tuning and continuously adjustable optical zoom. Opt. Express **2012**, 20, 27222–27229.

44. Tsou, Y.-S.; Chang, K.-H.; Lin, Y.-H. A droplet manipulation on a liquid crystal and polymer composite film as a concentrator and a sun tracker for a concentrating photovoltaic system. J. Appl. Phys. **2013**, 113, 244504.

45. Hassanfiroozi, A.; Huang, Y.-P.; Javidi, B.; Shieh, H.-P.D. Hexagonal liquid crystal lens array for 3D endoscopy. Opt. Express **2015**, 23, 971–981.

46. Asatryan, K.; Presnyakov, V.; Tork, A.; Zohrabyan, A.; Bagramyan, A.; Galstian, T. Optical lens with electrically variable focus using an optically hidden dielectric structure. Opt. Express **2010**, 18, 13981–13992.

47. Blum, R.D.; Kokonaski, W. Electro-active opthalmic lens having an optical power blending region. U.S. Patent 20,110,037,946, 17 Feruary 2011.

48. Lin, Y.-H.; Chen, H.-S. Electrically tunable-focusing and polarizer-free liquid crystal lenses for ophthalmic applications. Opt. Express **2013**, 21, 9428–9436.

49. Patel, J.S.; Rastani, K. Electrically controlled polarization-independent liquid-crystal Fresnel lens arrays. Opt. Lett. **1991**, 16, 532–534.

50. Ren, H.; Wu, S.-T. Adaptive liquid crystal lens with large focal length tunability. Opt. Express **2006**, 14, 11292–11298.

51. Harrold, J.; Wilkes, D.J.; Woodgate, G.J. Switchable 2D/3D display–solid phase liquid crystal microlens array. In Proceedings of the 11th International Display Workshops, Niigata, Japan, 8–10 December 2004.

52. Dai, H.T.; Liu, Y.J.; Sun, X.W.; Luo, D. A negative—positive tunable liquid-crystal microlens array by printing. Opt. Express **2009**, 17, 4317–4323.

53. Wang, B.; Ye, M.; Sato, S. Lens of electrically controllable focal length made by a glass lens and liquid-crystal layers. Appl. Opt. **2004**, 43, 3420–3425.

54. Commander, L.G.; Day, S.E.; Selviah, D.R. Variable focal length microlenses. Opt. Commun. **2000**, 177, 157–170.

55. Willemsen, O.H.; de Zwart, S.T.; IJzerman, W.L.; Hiddink, M.G.H.; Dekker, T. 2D/3D switchable displays. Proc. SPIE **2006**, 6135.

56. Flack, J.; Harrold, J.; Woodgate, G.J. A prototype 3D mobile phone equipped with a next-generation autostereoscopic display. Proc. SPIE **2007**, 6490.

57. Zhu, R.; Xu, S.; Hong, Q.; Wu, S.-T.; Lee, C.; Yang, C.-M.; Lo, C.-C.; Lien, A. Polymeric-lens-embedded 2D/3D switchable display with dramatically reduced crosstalk. Appl. Opt. **2014**, 53, 1388–1395.

58. Lin, Y.-H.; Ren, H.; Fan-Chiang, K.-H.; Choi, W.-K.; Gauza, S.; Zhu, X.; Wu, S.-T. Tunable-Focus Cylindrical Liquid Crystal Lenses. Jpn. J. Appl. Phys. **2005**, 44, 243–244.

59. Ye, M.; Wang, B.; Sato, S. Liquid-crystal lens with a focal length that is variable in a wide range. Appl. Opt. **2004**, 43, 6407–6412.

60. Kawamura, M.; Nakamura, K.; Sato, S. Liquid-crystal micro-lens array with two-divided and tetragonally hole-patterned electrodes. Opt. Express **2013**, 21, 26520–26526.

61. Kawamura, M.; Goto, H. Liquid crystal lens with two divided and double circularly hole-patterned electrodes. In Proceedings of the 2010 International Symposium on Optomechatronic Technologies, Toronto, ON, Canada, 5–27 October 2010; pp. 1–5.

62. Li, Y.; Wu, S.-T. Polarization independent adaptive microlens with a blue-phase liquid crystal. Opt. Express **2011**, 19, 8045–8050.

63. Hsu, C.-J.; Liao, C.-H.; Chen, B.-L.; Chih, S.-Y.; Huang, C.-Y. Polarization-insensitive liquid crystal microlens array with dual focal modes. Opt. Express **2014**, 22, 25925–25930.

64. Hsu, C.J.; Sheu, C.R. Using photopolymerization to achieve tunable liquid crystal lenses with coaxial bifocals. Opt. Express **2012**, 20, 4738–4746.

65. Hsu, C.-J.; Chih, S.-Y.; Jhang, J.-J.; Liao, C.-H.; Huang, C.-Y. Coaxially bifocal liquid crystal lens with switchable optical aperture. Liq. Cryst. **2015**, 1–7.

66. Zhao, X.; Liu, C.; Zhang, D.; Luo, Y. Tunable liquid crystal microlens array using hole patterned electrode structure with ultrathin glass slab. Appl. Opt. **2012**, 51, 3024–3030.

67. Huang, Y.-P.; Chen, C.-W.; Shen, T.-C.; Huang, J.-F. Autostereoscopic 3D display with scanning Multi-Electrode driven liquid crystal (MeD-LC) Lens. 3D Res. **2011**, 1, 39–42.

68. Li, K.; Robertson, B.; Pivnenko, M.; Deng, Y.; Chu, D.; Zhou, J.; Yao, J. High quality micro liquid crystal phase lenses for full resolution image steering in auto-stereoscopic displays. Opt. Express **2014**, 22, 21679–21689.

69. Urruchi Del Pozo, V.; Algorri Genaro, J.; Sánchez-Pena, J.; Geday, M.; Arregui, X.; Bennis, N. Lenticular arrays based on liquid crystals. Opto-Electronics Rev. **2012**, 20, 260–266.

70. Algorri, J.F.; Urruchi, V.; Bennis, N.; Sánchez-Pena, J.M. Cylindrical liquid crystal microlens array with rotary axis and tunable capability. IEEE Electron. Device Lett. **2015**, 36, 582–584.

71. Urruchi, V.; Algorri, J.F.; Marcos, C.; Sánchez-Pena, J.M. Note: Electrical modeling and characterization of voltage gradient in liquid crystal microlenses. Rev. Sci. Instrum. **2013**, 84, 116105.

72. Abramochkin, E.G.; Vasiliev, A.A.; Vashurin, P.V.; Zhmurova, L.I.; Ignatov, V.A.; Naumov, A.F. Controlled liquid crystal lens. Prepr. PN Lebedev Phys. Inst. **1988**, 194, 18.

73. Love, G.D.; Naumov, A.F. Modal liquid crystal lenses. Liq. Cryst. Today **2000**, 10, 1–4.

74. Fraval, N.; de La Tocnaye, J.L.D.B. Low aberrations symmetrical adaptive modal liquid crystal lens with short focal lengths. Appl. Opt. **2010**, 49, 2778–2783.

75. Loktev, M.Y.; Belopukhov, V.N.; Vladimirov, F.L.; Vdovin, G.V.; Love, G.D.; Naumov, A.F. Wave front control systems based on modal liquid crystal lenses. Rev. Sci. Instrum. **2000**, 71, 3290–3297.

76. Kirby, A.K.; Hands, P.J.; Love, G.D. Liquid crystal multi-mode lenses and axicons based on electronic phase shift control. Opt. Express **2007**, 15, 13496–13501.

77. Naumov, A.F.; Love, G.D.; Loktev, M.Y.; Vladimirov, F.L. Control optimization of spherical modal liquid crystal lenses. Opt. Express **1999**, 4, 344–352.

78. Hands, P.J.W.; Kirby, A.K. Adaptive modally addressed liquid crystal lenses. Proc. SPIE **2004**, 5518.

79. Kotova, S.P.; Patlan, V.V.; Samagin, S.A.; Zayakin, O.A. Wavefront formation using modal liquid-crystal correctors. Phys. Wave Phenom. **2010**, 18, 96–104.

80. Flexible Optical LC Lenses. Available online: http://www.okotech.com/lc (accessed on 6 January 2016).

81. Chang, Y.-C.; Jen, T.-H.; Ting, C.-H.; Huang, Y.-P. High-resistance liquid-crystal lens array for rotatable 2D/3D autostereoscopic display. Opt. Express **2014**, 22, 2714–2724.

82. Huang, Y.-P.; Chang, Y.-C.; Chen, C.-W.; Liao, L.-Y.; Shieh, P.-T.; Jen, T.-H.; Chen, T.-H. Adaptive Liquid Crystal Lens(LC Lens) Array for 3D Display and Capturing. In Proceedings of the Digital Holography and Three-Dimensional Imaging 2012, Miami, FL, USA, 28 April–2 May 2012.

83. Liu, Y.; Ren, H.; Xu, S.; Li, Y.; Wu, S.-T. Fast-response liquid-crystal lens for 3D displays. Proc. SPIE **2014**, 9005.

84. Urruchi, V.; Algorri, J.F.; Sánchez-Pena, J.M.; Bennis, N.; Quintana, X.; Otón, J.M. Electrooptic characterization of tunable cylindrical liquid crystal lenses. Mol. Cryst. Liq. Cryst. **2012**, 553, 211–219.

85. Khakifirooz, A.; Haji, S. UV-assisted nickel-induced crystallization of amorphous silicon. Thin Solid Films **2001**, 383, 241–243.

86. Mohiddon, M.A.; Lakshun Naidu, K.; Ghanashyam Krishna, M.; Dalba, G.; Rocca, F. Growth, optical, and electrical properties of silicon films produced by the metal-induced crystallization process. J. Nanopart. Res. **2011**, 13, 5999–6004.

CHAPTER 3

Structure, Photochromism and Liquid Crystal Properties of 1-Alkyl-2-(Arylazo)Imidazoles (Raai-C_nH_{2n+1}, n (Even) = 10 - 22)

Avijit Nandi[1], Chandana Sen[1], Debashis Mallick[2], Randhir K. Sinha[1], Chittaranjan Sinha[1*]

[1]Department of Chemistry, Inorganic Chemistry Section, Jadavpur University, Kolkata, India

[2]Department of Chemistry, Mrinalini Datta Mahavidyapith, Birati, India

ABSTRACT

Photochromism is the reversible structural change of the molecules upon light irradiation having different spectral pattern before and after irradiation. Liquid crystal (LC) is a long-range orientational order of the molecules in between solid and isotropic liquid phases which share the properties of the crystal and liquid phases. 1-Alkyl-2-(arylazo)imidazoles, (Raai-C_nH_{2n+1}, R = H, Me and n = 10, 12, 14, 16, 18, 20, 22) exhibits photochromism and two of them (Haai-$C_{18}H_{37}$ and Haai-$C_{22}H_{45}$) show Liquid crystal (LC) properties. One of these compounds, Meaai-$C_{16}H_{33}$ is structurally characterized by single crystal X-ray diffraction study. The compounds upon irradiation with UV light show E-to-Z isomerisation. Quantum yields ($f_{E®Z}$) of E-to-Z isomerisation vary with the molar mass of the compounds. The reverse transformation, Z-to-E, is recorded in thermal condition. The activation energy (E_a) of isomerisation is determined by the controlled temperature reaction. Liquid crystal property of the molecules is examined by polarizing optical microscopy (POM) and DSC. The carbon chain length n = 18 shows nematic (N)-isotropic (I) phase transition and an unknown Smectic (Sm) phase is observed with n = 22 on heating cycle.

KEYWORDS

Arylazoimidazoles; Structure; Photochromism; Liquid Crystal; Nematic-Isotropic Phase Transition; Smectic Phase

1. INTRODUCTION

Smart molecules can respond to a specific external stimulus in a specific manner and are of increasing interest in areas as diverse as microbiology [1], microfluidics [2], organic electronics [3,4], high density optical data storage devices [5,6] etc. Using of light as external stimulating agent has manifold advantages. Fritzsche (1867) observed the bleaching of an orange colored tetracene solution in daylight and then re-colored at night [7]. However, during the 1950s, the research groups of Hirshberg and Fischer carried out comprehensive studies on light irradiated color changes and the subsequent physical and chemical changes of different organic molecules [8]. The research field has received a renewable thrust in the 1980s when the photo-induced reversible transformation was observed for organic molecules like spirooxazines, diarylethenes etc. without photodegradation. This process is known as photochromism. It is defined as "reversible transformation of a chemical species induced in one or both directions by the absorption of electromagnetic radiation between two forms having different absorption spectra" [9]. The photochromic compounds have become appealing for their potential advantages in the construction of photo-switching devices and supramolecular architectures [9-15]. Various organic compounds show cis(Z)/trans(E) isomerization, pericyclic reactions, H atom transfer reactions, and dissociation processes. The molecules have been used as photoresponsive functional devices utilized as smart polymers [16], liquid crystals [17], molecular switches [18] and machines [19].

Azobenzene is used for a wide range of applications, from dye indicators and nonlinear optical chromophores to quantum interferences [20]. The molecule shows photoinduced E/Z isomerization by UV or visible light. More recently the synthesis and the characterization of heterocyclic azo dyes such as arylazopyridines have been extensively reported [21-24]. The prospect of azobenzene has encouraged us for synthesizing arylazoimidazoles [25] which acts as a potential switching group in biology and in coordination chemistry, since imidazole is a ubiquitous and essential group in biology, especially as a metal coordinating. The photochromism of 1-alkyl-2- (arylazo)imidazole (RaaiR) [26,27] and some of their complexes [28-31] are also reported.

The liquid crystal (LC) is a long-range orientational order and lies in between solid and isotropic liquid phases. The LC phase (mesophase) shares properties of the crystal and liquid phases. Some examples of anisotropic properties are birefringence, alignment in electric and magnetic fields (electrical permittivity and magnetic susceptibility), elasticity, viscosity, and conductivity [32,33]. Liquid crystals are widely used in information display technology and the anisotropic fluid states of rigid polymers are used for processing of high strength fibers. Liquid crystals (LCs) should also be useful in bio-related fields because the self-organized structures of lyotropic liquid crystals are compatible with those in living systems [34].

Comparing with azobenzene photochrome, arylazoimidazoles have several advantages of possessing both the photoactivity related to the azo unit, the capability of self-assembly through the imidazolyl group and coordinating ability to metal ions. Azoimidazoles display surprisingly a strong tendency toward self-assembly, which offers a new and efficient way to make photoactive liquid crystal (LC) materials. Attachment of long chain alkyl groups to aromatic molecules may develop liquid crystal (LC) property. The E form of azo derivatives is a rod and is nematic LCs (NLCs) while the Z is bent and tends to destabilize the phase structure. Therefore, E/Z photoisomerization of the azobenzene in the N phase can disorganize the phase structure of NLCs, resulting in a nematic-isotropic (N-I) isothermal phase transition (photochemical phase transition). In this article, we describe a preliminary account on photochromic LC (PLC) activity of some long chain alkyl group substituted 1-alkyl-2- (arylazo)imidazoles, Raai-C_nH_{2n+1} (n = 10, 12, 14, 16, 18, 20, 22).

2. EXPERIMENTAL

2.1. Materials

1-Alkyl-2-(arylazo)imidazoles were synthesized by reported procedure [27]. 1-Bromo-n-alkanes, n-$C_{10}H_{21}$-1- Br, n-$C_{12}H_{25}$-1-Br, $C_{14}H_{29}$-1-Br and 1-chloro-n-alkanes, $C_{16}H_{33}$-1-Cl, $C_{18}H_{37}$-1-Cl, $C_{20}H_{41}$-1-Cl and $C_{22}H_{45}$-1-Cl were purchased from Sigma-Aldrich and used as such. All other chemicals and solvents were analytical reagent grade as received.

2.2. Physical Measurements

Microanalytical data (C, H, N) were collected on PerkinElmer 2400 CHNS/O elemental analyzer. Spectroscopic data were obtained using the following instruments: UV-Vis spectra from a Perkin Elmer Lambda 25 spectrophotometer; IR spectra (KBr disk, 4000 - 400 cm^{-1}) from a Perkin Elmer RX-1 FTIR spectrophotometer; photo excitation has been carried out using a Perkin Elmer LS-55 spectrofluorimeter and ^{1}H NMR spectra were recorded from a Bruker (AC) 300 MHz FTNMR spectrometer. The liquid crystalline properties were established by thermal microscopy (Nikon polarizing microscope LV100POL attached with Linkam Temperature controlled stage model Examina-THMS600, and the phase transitions were confirmed by differential scanning calorimetry (Perkin-Elmer Diamond DSC Pyris1 system).

2.3. Synthesis of Compounds

2-(Arylazo)imidazoles were synthesized by the diazotization of aniline or p-toluidine in $NaNO_2/HCl$ at low temperature (0°C - 5°C) and followed by coupling with imidazole in aqueous sodium carbonate solution (pH-7). The alkylation was carried out by adding alkyl halide in dry THF solution to the corresponding 1-alkyl-2-(arylazo) imidazole in presence of NaH (Scheme 1). Long chain alkyl groups are appended at imidazole motif to synthesise present series of molecules. The compounds are 1-alkyl-2-(arylazo)imidazole [Raai-C_nH_{2n+1}, where R = H (a), Me (b); C_nH_{2n+1}—where n = 10 ($C_{10}H_{21}$, 1), 12 ($C_{12}H_{25}$, 2), 14 ($C_{14}H_{29}$, 3), 16 ($C_{16}H_{33}$, 4), 18 ($C_{18}H_{37}$, 5), 20 ($C_{20}H_{41}$, 6), 22 ($C_{22}H_{45}$, 7)]. Detail synthesis for Meaai-$C_{10}H_{21}$ (1b) is given below.

To dry THF solution (75 ml) of 2-(p-tolylazo)imidazole (2 g, 10.8 mmol), NaH (50% paraffin) (0.56 g) was added in small portion and stirred at cold condition on ice bath for 0.5 h. 1-Bromodecane (n-$C_{10}H_{21}$-1-Br; 2.50 g, 11.31 mmol) in dry THF (25 ml) was added slowly through pressure equalizing system under stirring condition for a period of 1 h and then warm for another 1 h on steam bath. The solution was evaporated to dryness, extracted with CH_2Cl_2, washed with NaOH solution (10%, 10 ml × 3) and finally by distilled water (20 ml × 3). The CH_2Cl_2 extract was chromatographed over silica gelcolumn (45 × 1 ml) prepared in petroleum ether (40°C - 60°C). The elution was performed by ethylacetate-petroleum ether (1:3 v/v) mixture and evaporated under reduced pressure. An orange crystalline product was obtained and then dried over P_4O_{10} under vacuum. Yield of the product was 2.28 g (65%).

All other compounds were prepared and purified by identical procedure and the yield was varied 55% - 70%. The microanalytical data of the compounds are as follows: Haai-$C_{10}H_{21}$ (1a), 54°C ± 1°C, Anal. Found: C, 73.00; H, 9.00; N, 18.00 Calc. for $C_{19}H_{28}N_4$: C, 73.08; H, 8.97; N, 17.95. Meaai-$C_{10}H_{21}$ (1b), 53°C ± 1°C, Anal. Found: C, 73.70; H, 9.13; N, 17.23 Calc. for $C_{20}H_{30}N_4$: C, 73.62; H, 9.20; N, 17.18. Haai-$C_{12}H_{25}$ (2a), 54°C ± 1°C, Anal. Found: C, 74.20; H, 9.36; N, 16.40. Calc. for $C_{21}H_{32}N_4$: C, 74.12; H, 9.41; N, 16.47. Meaai-$C_{12}H_{25}$ (2b), 52°C ± 1°C, Anal. Found: C, 74.64; H, 9.55; N, 15.90 Calc. for $C_{22}H_{34}N_4$: C, 74.58; H, 9.60; N, 15.82. Haai-$C_{14}H_{29}$ (3a), 54°C ± 1°C, Anal. Found: C, 74.94; H, 9.84; N, 15.30 Calc. for $C_{23}H_{36}N_4$: C, 75.00; H, 9.78; N, 15.22. Meaai-$C_{14}H_{29}$ (3b), 53°C ± 1°C Anal. Found: C, 75.47; H, 9.92; N, 14.72 Calc. for $C_{24}H_{38}N_4$: C, 75.39; H, 9.95; N, 14.66. Haai-$C_{16}H_{33}$ (4a), 53 ± 1°C, Anal. Found: C, 75.70; H, 10.05; N, 14.20 Calc. for $C_{25}H_{40}N_4$: C, 75.76; H, 10.10; N, 14.14. Meaai-$C_{16}H_{33}$ (4b), 54°C ± 1°C, Anal. Found: C, 76.12; H, 10.20; N, 13.60 Calc. for $C_{26}H_{42}N_4$: C, 76.09; H, 10.24; N, 13.66. Haai-$C_{18}H_{37}$ (5a), 59°C ± 1°C, Anal.

Found: C, 76.50; H, 10.35; N, 13.30. Calc. for $C_{27}H_{44}N_4$: C, 76.42; H, 10.38; N, 13.21. Meaai$C_{18}H_{37}$ (5b), 56°C ± 1°C, Anal. Found: C, 76.80; H, 10.55; N, 12.72 Calc. for $C_{28}H_{46}N_4$: C, 76.71; H, 10.50; N, 12.79. Haai-$C_{20}H_{41}$ (6a), 65°C ± 1°C, Anal. Found: C, 76.94; H, 10.66; N, 12.30 Calc. for $C_{29}H_{48}N_4$: C, 76.99; H, 10.62; N, 12.39. Meaai-$C_{20}H_{41}$ (6b), 63°C ± 1°C, Anal. Found: C, 77.31; H, 10.76; N, 12.06 Calc. for $C_{30}H_{50}N_4$: C, 77.25; H, 10.73; N, 12.02. Haai-$C_{22}H_{45}$ (7a), 65°C ± 1°C, Anal. Found: C, 77.60; H, 10.86; N, 11.70 Calc. for $C_{31}H_{52}N_4$: C, 77.51; H, 10.83; N, 11.67. Meaai-$C_{22}H_{45}$ (7b), 63°C ± 1°C, Anal. Found: C, 77.80 H, 10.88; N, 11.30 Calc. for $C_{32}H_{54}N_4$: C, 77.73; H, 10.93; N, 11.34%.

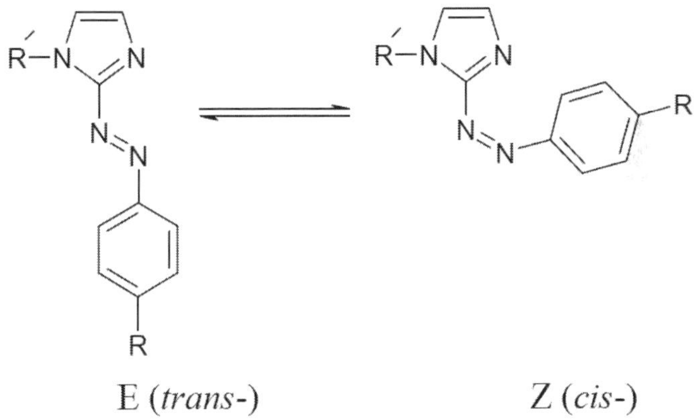

Scheme 1. Isomerization of 1-alkyl-2-(arylazo)imidazole.

2.4. X-Ray Diffraction Study

The crystallographic data are shown in Table 1. Suitable single crystal of Meaai-$C_{16}H_{33}$ (4b) was mounted on a Siemens CCD diffractometer equipped with graphite monochromated Mo-K_a (l = 0.71073 Å) radiation. The unit cell parameters and crystal-orientation matrices were determined for two complexes by least squares refinements of all reflections. The intensity data were corrected for Lorentz and polarisation effects and an empirical absorption correction were also employed using the SAINT program. Data were collected applying the condition I > 2s(I). All these structures were solved by direct methods and followed by successive Fourier and difference Fourier syntheses. Full matrix least squares refinements on F^2 were carried out using SHELXL-97 with anisotropic displacement parameters for all non-hydrogen atoms. Hydrogen atoms were constrained to ride on the respective carbon or nitrogen atoms with an isotropic displacement parameters equal to 1.2 times the equivalent isotropic displacement of their parent atom in all cases. Complex neutral atom scattering factors were used throughout for all cases. All calculations were carried out using SHELXS 97 [35], SHELXL97 [36] and PLATON 99 [37].

2.5. Computational Study

The direction of the dipole of the compound 5a was calculated by Density Functional Theory (DFT), which may determine the presence of dipolar interaction within the molecules. All calculations have been performed by the Chem3D (version 10) [38] with GAUSSIAN 03 Interface [39]. The first optimization was done by the MM2 method, and finally DFT was performed as stated earlier.

3. RESULTS AND DISCUSSION

3.1. Synthesis and Formulation

1-Alkyl-2-(arylazo)imidazole (Raai-C_nH_{2n+1}) are synthesized by controlled alkylation of 2-(arylazo)imidazole (RaaiH) by adding alkyl halide ($C_nH_{2n+1}X$) in dry THF using NaH. RaaiH are synthesized by diazo reaction of aryldiazonium salt with imidazole at pH 7 (Scheme 2). Alkyl halides of varying carbon chain (C_{10} to C_{22}) are used to synthesize Raai-C_nH_{2n+1} (n = 10 ($C_{10}H_{21}$, 1), 12 ($C_{12}H_{25}$, 2), 14 ($C_{14}H_{29}$, 3), 16 ($C_{16}H_{33}$, 4), 18 ($C_{18}H_{37}$, 5), 20 ($C_{20}H_{41}$, 6), 22 ($C_{22}H_{45}$, 7)). The compounds are purified by chromatographic separation in silica gel column using acetonitrile-benzene (8:2, v/v). The elemental composition is supported by microanalytical data and they are low melting solid (vide Experimental section).

3.2. Spectral Studies

A moderately intense stretching for n(C=N) is observed at 1585 - 1615 and n(N=N) at 1370 - 1374 cm^{-1} (Table 2).

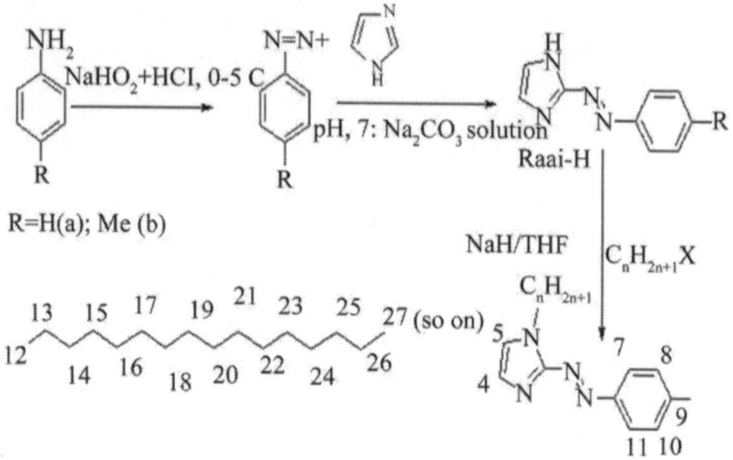

Scheme 2. Synthetic process of Raai-C_nH_{2n+1} (1 - 7).

C_nH_{2n+1} n = 10 ($C_{10}H_{21}$, 1), 12 ($C_{12}H_{25}$, 2), 14 ($C_{14}H_{29}$, 3), 16 ($C_{16}H_{33}$, 4), 18 ($C_{18}H_{37}$, 5), 20 ($C_{20}H_{41}$, 6), 22 ($C_{22}H_{45}$, 7)

Table 1. Summarized crystallographic data for Meaai$C_{16}H_{33}$ (4b).

	4b
Empirical formula	$C_{26}H_{42}N_4$
Formula weight	410.64
Temperature (K)	296
Crystal system	Monoclinic
Space group	P21/c
Crystal size $(mm)^3$	0.10 × 0.35 × 0.40
a(Å)	24.405(6)
b(Å)	11.049(3)
c(Å)	9.927(3)
β(°)	93.179(7)
$V(Å)^3$	2672.7(13)
Z	4
μ ($MoK_α$) (mm^{-1})	0.060
θ range	1.7 - 25.1
hkl range	−29 < h < 28; −13 < k < 12; −8 < l < 11
D_{calc} ($mg{\cdot}m^{-3}$)	1.020
Refine parameters	273
Total reflections	15,392
Unique reflections	4696
R_1[a] [I > 2σ(I)]	0.0647
wR_2[b]	0.1409
Goodness of fit	1.039

[a]$R = \Sigma||F_o|-|F_c||/\Sigma|F_o|$; [b]$wR_2 = [\Sigma w(F_o^2-F_c^2)^2/\Sigma w(F_o^2)^2]^{1/2}$, $w = 1/[\sigma^2(F_0)^2 + (0.0986P)^2]$ for 4b; where $P = (F_0^2 2^2 + 2Fc^2 2^2)/3$.

The stretching frequency of –N=N- (exocylic) are insignificantly affected on comparing with –C=N- (imidazolyl) group. This may support effect of alkylation at N(1)-imidazolyl position of the compounds [27].

The absorption spectra of the compounds were recorded in MeOH solution in the wavelength range 200 - 900 nm (Figure 1, Table 2). The absorption spectra of Raai-C_nH_{2n+1} show main absorption band at 355 - 375 nm with a molar absorption coefficient in the order of 10^4 $M^{-1}\cdot cm^2$ and a weak band at 260 - 280 nm and 450 - 455 nm. The intense absorption band corresponds to p-p* transitions, while the tail portion is due to n-p* transition. With increasing chain length the absorption maxima have been shifted to longer wavelength which supports the enhancement of inductive effect of substituent with increase in number of bonded carbon centres at N(1)-imidazolyl group. The –N=Nis a part of a conjugated system of double bonds, both the n→π* and the π→π* bands are shifted to longer wavelength [40].

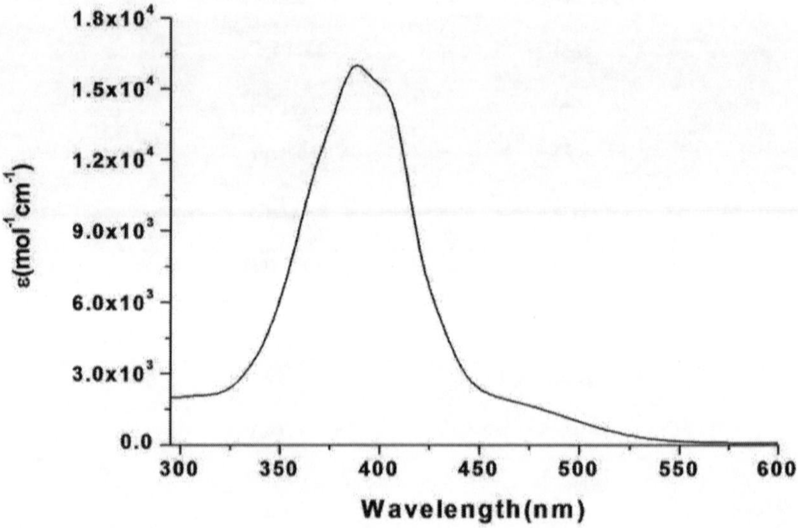

Figure 1. UV-Vis spectrum of Haai-$C_{20}H_{41}$ (6a) in methanol.

The ^{1}H NMR spectra of Raai-C_nH_{2n+1} (1-7) are recorded in $CDCl_3$. The atom numbering is shown in Scheme 2. The spectral data are set out in Table 3. The proton signals are assigned on the basis of spin-spin interaction and effect of substitution. The alkylation of imidazole is supported by the disappearance of d(N–H) at ~10.30 ppm and the appearance of 1-alkyl signal at 0.85 - 4.40 ppm for -N–CH_2–(CH_2)n–CH_3. A triplet signal for -N–CH_2– appears at 4.40 ppm; and multiplet is observed for –(CH_2)n– at 0.80 - 1.50 ppm. Imidazolyl 4- and 5-H appear as broad singlet at 7.24 - 7.28 and 7.14 - 7.17 ppm, respectively. Broadening may be due to rapid proton exchange between these imidazolic protons. The aryl protons (7-H to 11-H) are upfield shifted on going from phenylazo (a) to p-tolylazo (b) which may be due to +I effect of –Me group.

3.3. Molecular Structure

The molecular structure of Meaai-$C_{16}H_{33}$ (4b) is shown in Figure 2. The bond parameters are listed in Table 4" target="_self"> Table 4. The bond length and angles of the fragment of the structure, (p-tolylazo)imidazole, are comparable to the reported structure of Haai-Me [27]. The azo N-N distance is 1.266(4) Å which is slightly longer than the reported value (1.257(6) Å) [27]. Two groups around N=N are not perfectly planar and the dihedral angle between the two least squares planes is avg. 32°. The crowding of long chain, 1-$C_{16}H_{33}$ may be the reason for deviation. The bonding strength between N and p-tolyl-C [C(8)–N(2), 1.421(4) Å] is weaker than that of N and imidazolyl-C [C(1)–N(1), 1.385(4) Å]. This suggests that the intramolecular interaction between azo-N and imidazolyl-C wave functions is stronger than that of p-tolyl orbitals. The C(3)-C(4) bond length (1.364(5) Å) of imidazolyl group is somewhat greater than reported data [41] which may be due to the steric demand of 1-$C_{16}H_{33}$ appended to imidazolyl group. The bond angle values at chain of C(15)-C(16)-C(17) (116.3(4)°) C(20)-C(21)-C(22) (120.6 (6)°), C(24)-C(25)-C(26) (123.6(8)°) etc. show gradual increment which may be due to the removal of steric strain from the neighboring groups arranged in a zigzag manner. The imidazolyl-N is hydrogen bonded, N(3)-- -H(3)-C(3), (N(3)---H(3), 2.51 Å; N(3)---C(3), 3.390(5) Å and Ð N(3)---H(3)-C(3), 158°; symmetry, -x, -1/2+y, 1/2-z) to imidazolyl-C-H giving 1D chain (Figure 3). Packing view in the unit cell shows π---π interaction (3.3477(16) Å; symmetry, 1-x, 2-y, -z) between adjacent imidazolyl groups to form 2D supramolecular structure (Figure 4).

3.4. Photochromism

The irradiation of UV light at l_{max} to a MeOH solution of the ligands has shown the absorption spectral change (Figure 5). The intense peak at l_{max} decreases, which is accompanied by a slight increase at the tail portion of the spectrum until a photostationary state (PSS-I) is reached. Subsequent irradiation at the newly appeared longer wavelength peak reverses the course of the reaction and the original spectrum is recovered up to a point, which is another photostationary state (PSS-II) under irradiation at the longer wavelength peak. The quantum yields of the E-to-Z isomerisation were determined following previously reported procedure [26,27] and the results are set out in Table 5.

Table 2. FT-IR[a] and UV-VIS[b] spectral data of the compounds.

Compounds	IR in KBr disc (cm^{-1})		UV-Vis spectral data λ_{max}(nm), (10^{-3} ε ($dm^3 \cdot mol^{-1} \cdot cm^{-1}$))
	ν(N=N)	ν(C=N)	
Haai-$C_{10}H_{21}$ (1a)	1403	1624	362 (25.9), 374 (24.1), 451 (2.4)
Meaai-$C_{10}H_{21}$ (1b)	1404	1625	363 (25.9), 376 (22.9), 451(2.1)
Haai-$C_{12}H_{25}$ (2a)	1404	1624	363 (27.1), 372 (23.9), 451(2.3)
Meaai-$C_{12}H_{25}$ (2b)	1405	1623	364 (28.9), 376 (22.9) ,452 (2.3)
Haai-$C_{14}H_{29}$ (3a)	1403	1624	365 (25.2), 376 (24.2), 454 (2.1)
Meaai-$C_{14}H_{29}$ (3b)	1403	1625	364 (26.9), 375 (23.6), 455 (2.2)
Haai-$C_{16}H_{33}$ (4a)	1404	1624	365 (26.2), 375 (24.1), 451 (2.4)
Meaai-$C_{16}H_{33}$ (4b)	1404	1625	367 (25.9), 377 (24.5), 453 (2.1)
Haai-$C_{18}H_{37}$ (5a)	1405	1623	367 (25.8), 378 (24.2), 454 (2.1)
Meaai-$C_{18}H_{37}$ (5b)	1405	1625	368 (26.9), 378 (26.1), 455 (2.2)
Haai-$C_{20}H_{41}$ (6a)	1404	1624	370 (25.2), 380 (22.1), 458 (2.4)
Meaai-$C_{20}H_{41}$ (6b)	1405	1624	372 (25.9), 385 (23.6), 460 (2.1)
Haai-$C_{22}H_{45}$ (7a)	1404	1622	373 (19.3), 386 (17.3), 460 (1.98)
Meaai-$C_{22}H_{45}$ (7b)	1406	1620	373 (2.3), 385 (29.2), 462 (3.6)

[a]In KBr disc; [b]In methanol.

Table 3. ^{1}H NMR spectral data in $CDCl_3$ at room temperature.

Compounds	4-H^{s}	5-H^{s}	7,11-H^{d}	8,10-H	9-R^{s}	12-CH_2^{t}	$-(CH_2)_{n-1}-CH_3^{*}$
Haai-$C_{10}H_{21}$ (1a)	7.35	7.16	7.97 (8.2)	7.45^{m}	7.43	4.43 (6.7)	1.88 - 1.28
Meaai-$C_{10}H_{21}$ (1b)	7.32	7.15	7.88 (8.5)	7.28 (7.0)^{d}	2.42	4.40 (6.4)	1.87 - 1.29
Haai-$C_{12}H_{25}$ (2a)	7.34	7.17	7.94 (8.0)	7.46^{m}	7.43	4.44 (7.1)	1.88 - 1.34
Meaai-$C_{12}H_{25}$ (2b)	7.31	7.14	7.90 (7.5)	7.28 (7.0)^{d}	2.44	4.40 (6.4)	1.86 - 1.23
Haai-$C_{14}H_{29}$ (3a)	7.34	7.17	7.96 (8.1)	7.47^{m}	7.46	4.42 (6.6)	1.89 - 1.25
Meaai-$C_{14}H_{29}$ (3b)	7.33	7.14	7.89 (8.6)	7.27 (7.0)^{d}	2.41	4.41 (7.0)	1.89 - 1.28
Haai-$C_{16}H_{33}$ (4a)	7.35	7.16	7.97 (8.0)	7.49^{m}	7.49	4.42 (6.6)	1.90 - 1.25
Meaai-$C_{16}H_{33}$ (4b)	7.27	7.14	7.96 (7.8)	7.27 (7.4)^{d}	2.44	4.43 (7.0)	1.70 - 1.22
Haai-$C_{18}H_{37}$ (5a)	7.36	7.17	7.97 (8.2)	7.50^{m}	7.49	4.41 (7.0)	1.89 - 1.23
Meaai-$C_{18}H_{37}$ (5b)	7.31	7.14	7.88 (8.5)	7.26 (7.4)^{d}	2.43	4.40 (7.2)	1.87 - 1.21
Haai-$C_{20}H_{41}$ (6a)	7.35	7.16	7.96 (8.1)	7.52^{m}	7.51	4.41 (7.0)	1.88 - 1.24
Meaai-$C_{20}H_{21}$ (6b)	7.32	7.14	7.89 (8.6)	7.27 (7.0)^{d}	2.41	4.43 (7.0)	1.89 - 1.21
Haai-$C_{22}H_{45}$ (7a)	7.36	7.16	7.97 (8.2)	7.53^{m}	7.52	4.42 (6.6)	1.90 - 1.22
Meaai-$C_{22}H_{45}$ (7b)	7.27	7.14	7.88 (8.5)	7.26 (7.0)^{d}	2.41	4.40 (7.2)	1.86 - 1.22

^{s}Singlet; ^{d}Doublet; ^{t}Triplet; ^{bs}Broad Singlet; ^{m}Multiplet; *protons of $-(CH_2)_{n-1}-CH_3$.

Table 4.. Selected bond distances (Å) and angles (˚) for Meaai-$C_{16}H_{33}$ (4b).

Bond distances (Å)		Bond angles (˚)	
N(2)-C(1)	1.369(4)	C(12)-C(13)-C(14)	116.1(4)
N (2)-C(3)	1.352(50)	C(13)-C(14)-C(15)	117.4(4)
C(2)-C(3)	1.365(5)	C(23)-C(24)-C(25)	120.7(8)
N(1)-C(2)	1.356(4)	C(24)-C(25)-C(26)	119.8(9)
N(3)-C(1)	1.384(5)		
N(4-N(3)	1.262(4)		
N(4)-C(4)			

The absorption spectra of the E-isomer of ligands have changed with isosbestic points upon excitation (Figure 5) to the Z-isomer. The ligands show little sign of degradation upon repeated irradiation (Figure 5(b)). The quantum yields were measured for the E-to-Z ($f_{E®Z}$) photoisomerisation of these ligands in MeOH. The $f_{E®Z}$ values are significantly dependent on nature of substituents, and chain length of -C_nH_{2n+1}. The Me substituent, HaaiC_nH_{2n+1} to Meaai-C_nH_{2n+1}, and the alkyl substituent C_nH_{2n+1} (n = 10, 12, 14, 16, 18, 20, 22) at N(1)-position reduce $f_{E®Z}$ values. In general, increase in mass of the molecule reduces the rate of isomerisation, E-to-Z. The rotor volume and mass have significant influence on the isomerisation rate and quantum yields.

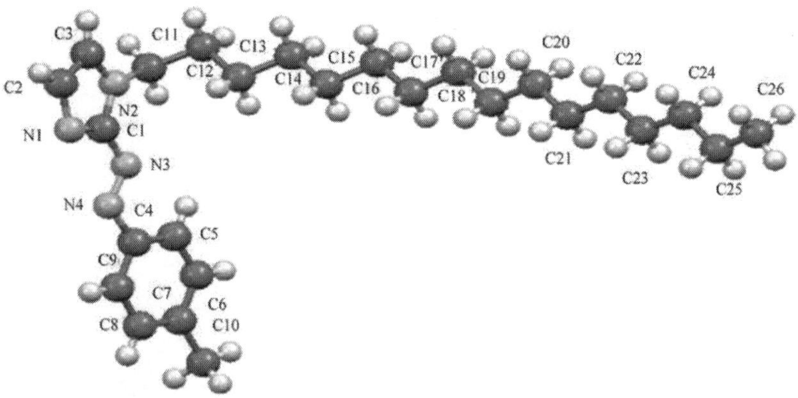

Figure 2. Crystal structure of Meaai-$C_{16}H_{33}$ (4b).

Thermal Z-to-E isomerisation of the ligands was followed by UV-Vis spectroscopy in MeOH at 298 - 313 K (Table 6). The Eyring plots in the range 298 - 313 K gave a linear graph from which the activation energy was obtained (Table 6, Figure 6). With the increase of the alkyl chain length on N atom of the imidazole moiety, the E_as are reduced which means faster Z-to-E thermal isomerisation. The entropy of activation (DS^*) are high negative in the organic compounds having large alkyl groups than that of the compounds having short alkyl chain. This is also in support of increase in rotor volume in the organic fragments.

3.5. Liquid Crystal Behaviour and Tansition Enthalpies of Haai-C_nH_{2n+1} (n = 18 (5a) and 22 (7a))

The thermal properties of the compounds were studied using polarized light optical microscopy (POM) and differential scanning calorimetry (DSC). Out of fourteen compounds only two compounds, Haai-$C_{18}H_{37}$ (5a) and Haai-$C_{22}H_{45}$ (7a) show mesomorphic response. The sample is sandwiched between untreated slide and a coverslip and is under hot stage of POM. Other twelve compounds show common crystalline-isotropic transition and the DSC plots do not show significant changes. The compound, 5a exhibits nematic (N) texture at 55°C (Figure 7) while 7a shows unknown Smectic (Sm) phase at 62°C (Figure 8) and, finally shift to isotropic (I) phase at 58°C (5a) and 65°C (7a) respectively. The crystal and nematic phases are under thermal equilibrium. During cooling from isotropic phase, 5a shows strongly fluctuating textures of the N phase and has been transformed to textures in which "Schlieren" regions can be seen (Figure 7). Homeotropic and homogenous oriented domains cause this coexistence, which permits the assignment of the smectic phase in compound 7a. The transition temperature and associated enthalpies are Cr → N 37.4°C (15.9 $KJ{\cdot}mol^{-1}$) → Iso 58°C (0.5 $KJ{\cdot}mol^{-1}$/heating cycle) → N 44.1°C (12.6 $KJ{\cdot}mol^{-1}$/cooling cycle) for 5a (Figure 9).

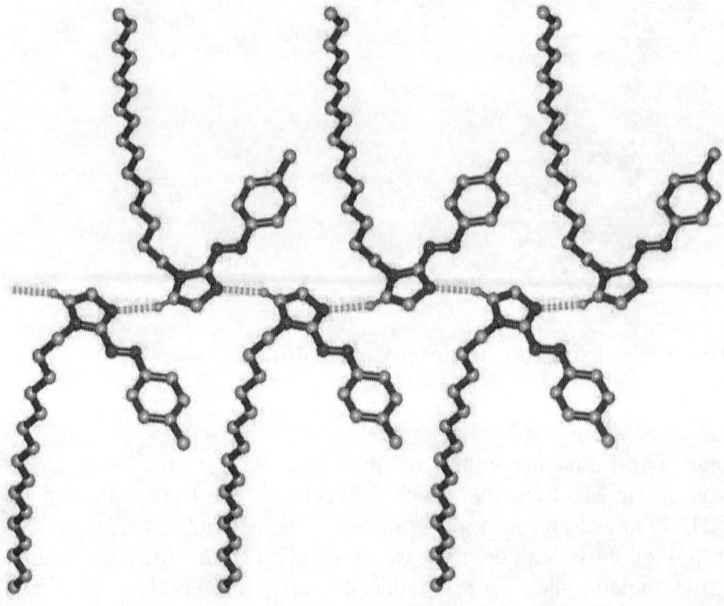

Figure 3. 1D supramolecular chains are formed along crystallographic b-axis through hydrogen bonding interactions of Meaai-$C_{16}H_{33}$ (4b).

3.6. Computational Study

The Density Functional Theory (DFT) computation has been used to calculate dipole moment of 5a and 7a. It may determine the presence of dipolar interaction within the molecules (Figure 10). The dipole moment of 5a is 2.997 Debye (X = −1.3726, Y = −2.4777, Z = 0.9814). After geometry optimization of the compound 5a by addition of atoms step by step, the optimized structure suggests that the molecule possess rod shape. The first optimization was done by the MM2 method, and finally DFT was performed. The maxima of dipole moment are in the negative Y-direction (−2.4777 D), so the direction of net dipole moment should be along the molecular axis (axial). Therefore, the direction of net dipole moment is axial (along molecular axis). Near-neighbor antiparallel correlation of molecules within the layer of N phase will occur in reality if the direction of dipole is axial or along the molecular axis. The direction of rod shaped compound, 5a in adjacent layers should be antiparallel, so that the layer polarization alternates from layer to layer; leading to a macroscopic apolar antiferroelectric structure which is applicable to N phases and sometime in case of reentrant Smectic phase with switching behavior. Within the layers the molecules are in most cases additionally non-tilted parallel to the polar direction in the case of the Nematic phase. DFT study suggests that the nearneighbour antiparallel correlation exists and in broad-on side position with respect to the direction of dipole within the layers. Based on the direction of the dipole moment, calculated from DFT we have constructed a model of supramolecular arrangement in Nematic phase which is depicted in Figure 11.

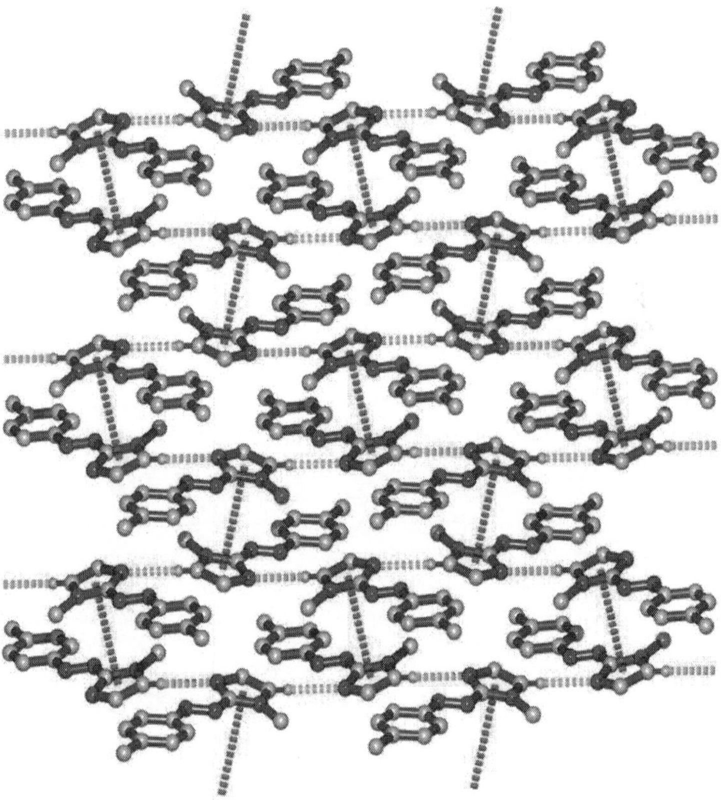

Figure 4. 2D supramolecular structure of Meaai-$C_{16}H_{33}$ (4b) formed by hydrogen bonding (Green) and π---π (magenta) interactions (for clarity the long aliphatic chains started from C15 (Red in color) and hydrogen atoms are not given).

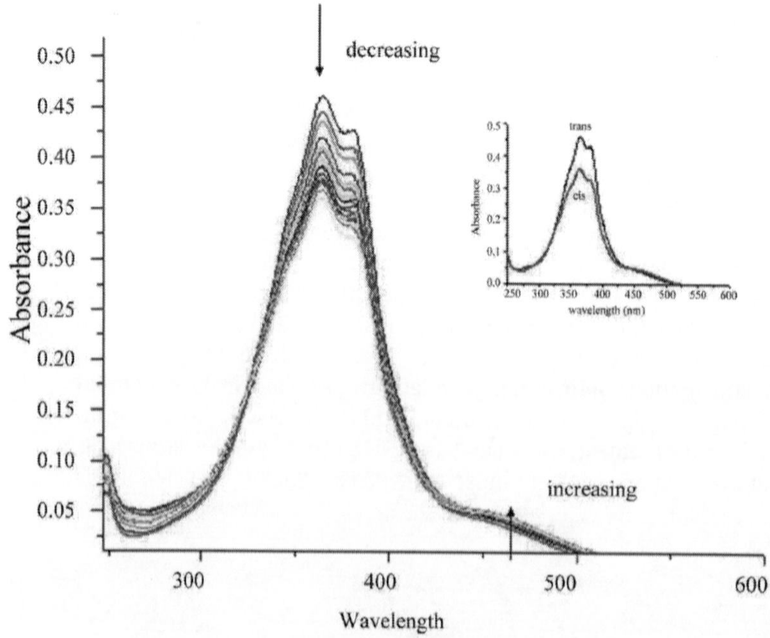

(a)

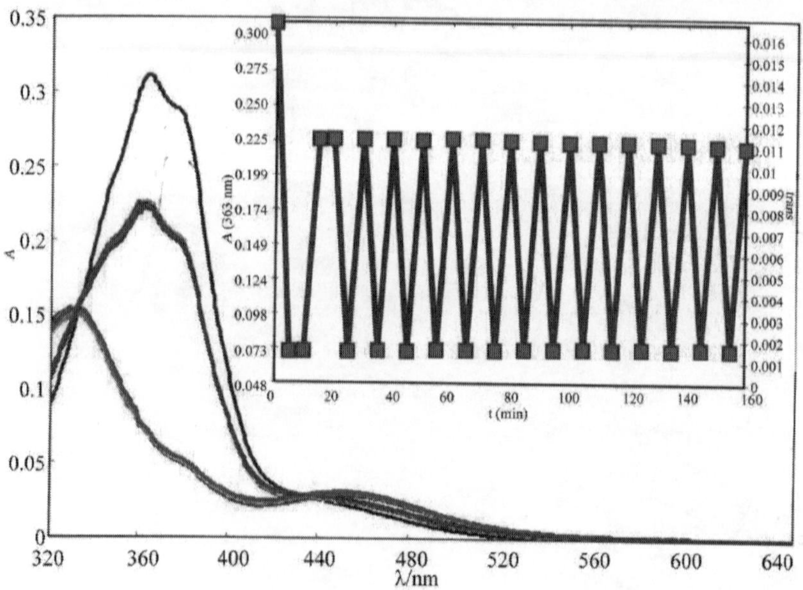

(b)

Figure 5. (a) Spectral changes of Haai-$C_{16}H_{33}$ (4a) in MeOH upon repeated irradiation at 365 nm at 3 min interval at 25°C. Inset figure shows the spectra of Z and E isomers of Haai-$C_{16}H_{33}$; (b) Resistance to repeated irradiation in MeOH solution of Haai-$C_{16}H_{33}$ (4a).

Table 5. Results of photochromism, rate of conversion and quantum yields upon UV light irradiation.

Compounds	λ_{π,π^*} (nm)	Isobestic point (nm)	Rate of E→Z conversion × 10^9 (s^{-1})	$\phi_{E\rightarrow Z}$ Conversion × 10^9
Haai-$C_{10}H_{21}$ (1a)	362	328, 424	34.11	1.61
Meaai-$C_{10}H_{21}$ (1b)	363	326, 425	33.12	1.43
Haai-$C_{12}H_{25}$ (2a)	363	330, 427	32.76	1.54
Meaai-$C_{12}H_{25}$ (2b)	364	334, 425	32.14	1.38
Haai-$C_{14}H_{29}$ (3a)	361	326, 426	28.31	1.47
Meaai-$C_{14}H_{29}$ (3b)	361	328, 425	26.14	1.35
Haai-$C_{16}H_{33}$ (4a)	360	323, 432	25.81	1.37
Meaai-$C_{16}H_{33}$ (4b)	362	318, 436	24.80	1.29
Haai-$C_{18}H_{37}$ (5a)	362	315, 435	23.24	1.27
Meaai-$C_{18}H_{37}$ (5b)	363	312, 432	22.71	1.24
Haai-$C_{20}H_{41}$ (6a)	361	318, 429	22.13	1.20
Meaai-$C_{20}H_{21}$ (6b)	362	313, 445	21.14	1.19
Haai-$C_{22}H_{45}$ (7a)	360	319, 433	20.72	1.17
Meaai-$C_{22}H_{45}$ (7b)	363	311, 446	20.17	1.11

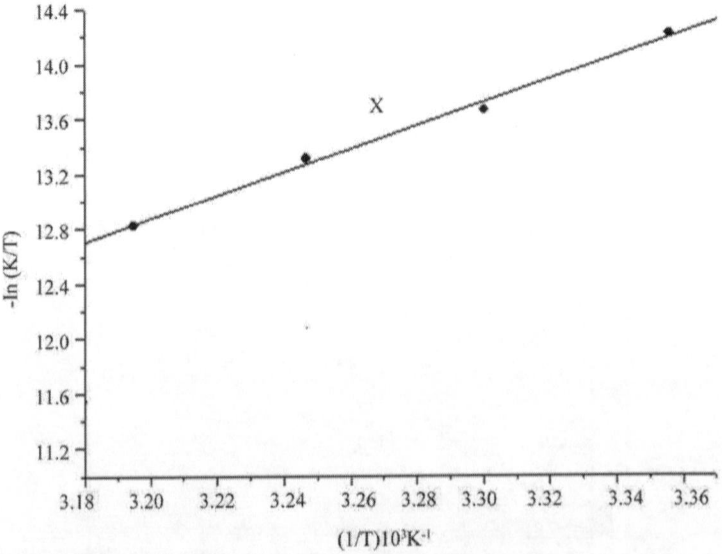

Figure 6. The Eyring plot of rate constants of Z-to-E thermal isomerisation of Haai-$C_{16}H_{33}$ (4a).

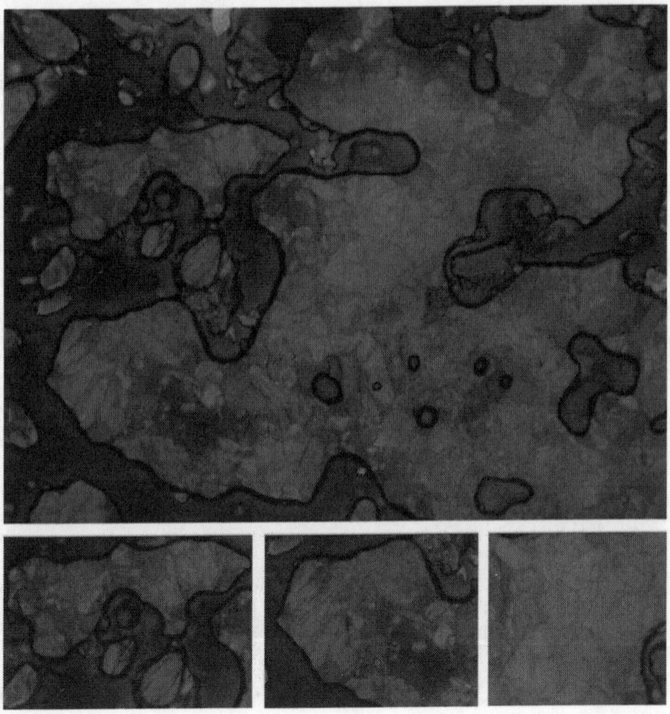

Figure 7. Fluctuating texture with schlieren texture of N phase at 51.2°C of Haai-$C_{18}H_{37}$ (5a).

4. CONCLUSION

Seven pairs of 1-alkyl-2-(arylazo)imidazoles, $RaaiC_nH_{2n+1}$ (n = 10, 12, 14, 16, 18, 20, 22) with long alkyl chain at imidazolyl-N(1) centre are spectroscopically characterised. As one of the compounds, 1-hexadcayl-2- (p-tolylazo)imidazole, Meaai-$C_{16}H_{33}$ is characterised by single crystal X-ray diffraction study. Upon UV light irradiation, the compounds show photochromism through geometrical isomerisation, E (trans)-to-Z (cis) forms. The reverse process is followed by thermal route and potential energy barrier (E_a) is calculated. The presence of long chain alkyl group shows long-range orientational order in between solid (crystalline, Cr) and isotropic (I) liquid phases. In some cases Cr and Nematic (N) texture shows thermal equilibrium in heating/cooling cycle. The Schlieren texture of N phase observed in compound 5a and 7a exhibits unknown smectic pahse with symmetrical lines. This is a preliminary report of our work; 1, 3-dialkyl-2-(arylazo)imida-zolium salts of different anions and the metal complexes (transition and nontransition) are waiting in this series.

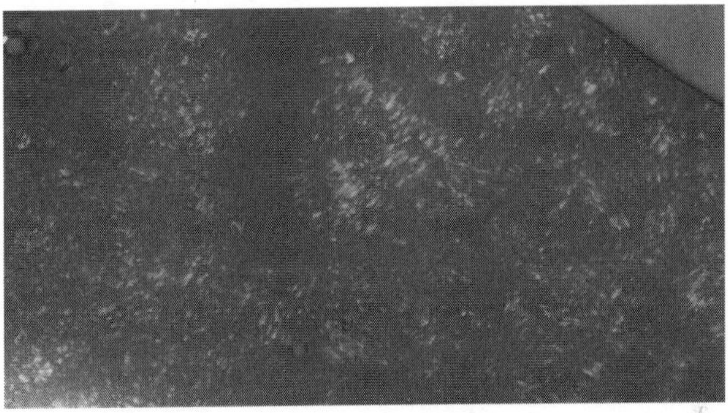

Figure 8. Unknown texture of Sm phase at 62°C of $HaaiC_{22}H_{43}$.

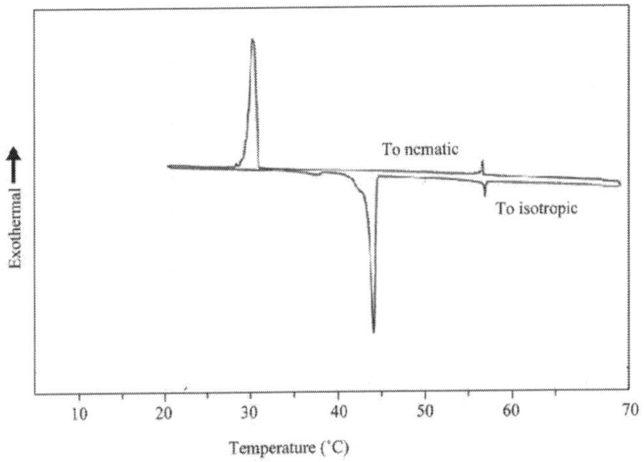

Figure 9. DSC of Haai-$C_{18}H_{37}$ (5a) at the rate of 2°C·min^{-1}.

Table 6. Rate and activation parameters for Z (cis) ® E (trans) thermal isomerisation.

Compounds	Temp (K)	Rate of thermal Z→E conversion × 10^3 (s^{-1})	E, kJ mol^{-1}	ΔH$^{\ne}$ kJ·mol^{-1}	ΔS$^{\ne}$ J·mol^{-1}·K^{-1}	ΔG$^{\ne}$ kJ·mol^{-1}
	298	0.2292				
Haai-$C_{10}H_{21}$ (1a)	303	0.3898	71.67	69.48	−81.94	92.14
	308	0.5496				
	313	0.8867				
	298	0.2345				
Meaai-$C_{10}H_{21}$ (1b)	303	0.3976	70.19	70.13	−83.47	90.12
	308	0.5634				
	313	0.8979				
	298	0.2367				
Haai-$C_{12}H_{25}$ (2a)	303	0.3997	69.86	68.97	−82.94	91.75
	308	0.5645				
	313	0.8997				
	298	0.2654				
Meaai-$C_{12}H_{25}$ (2b)	303	0.4235	69.06	68.89	−84.47	89.97
	308	0.6167				
	313	0.9546				
	298	0.2567				
Haai-$C_{14}H_{29}$ (3a)	303	0.4135	67.38	67.78	−84.49	90.34
	308	0.5896				
	313	0.9123				
	298	0.2657				
Meaai-$C_{14}H_{29}$ (3b)	303	0.4767	67.78	69.59	−86.67	86.67
	308	0.6578				
	313	0.9545				
	298	0.2789				
Haai-$C_{16}H_{33}$ (4a)	303	0.4567	65.34	67.73	−85.69	89.37
	308	0.6189				
	313	1.0989				
	298	0.2577				
Meaai-$C_{16}H_{33}$ (4b)	303	0.4247	73.40	70.85	−76.45	94.02
	308	0.6276				
	313	0.9571				
	298	0.3467				
Haai-$C_{18}H_{37}$ (5a)	303	0.5567	62.21	63.32	−87.79	87.79
	308	0.7567				
	313	1.4567				
	298	0.4534				
Meaai-$C_{18}H_{37}$ (5b)	303	0.6989	62.08	70.56	−89.97	85.54
	308	0.8998				
	313	1.9675				
	298	0.4569				
Haai-$C_{20}H_{41}$ (6a)	303	0.6789	60.45	61.89	−91.9	85.67
	308	0.8765				
	313	1.9876				
	298	0.6578				
Meaai-$C_{20}H_{41}$ (6b)	303	0.7568	60.09	69.99	−92.45	82.23
	308	0.9678				
	313	2.1434				
	298	0.4817				
Haai-$C_{22}H_{45}$ (7a)	303	0.7541	58.64	60.89	−92.38	83.45
	308	0.9574				
	313	2.4270				
	298	0.7140				
Meaai-$C_{22}H_{45}$ (7b)	303	0.9247	58.97	78.27	−43.90	91.56
	308	1.2613				
	313	3.729				

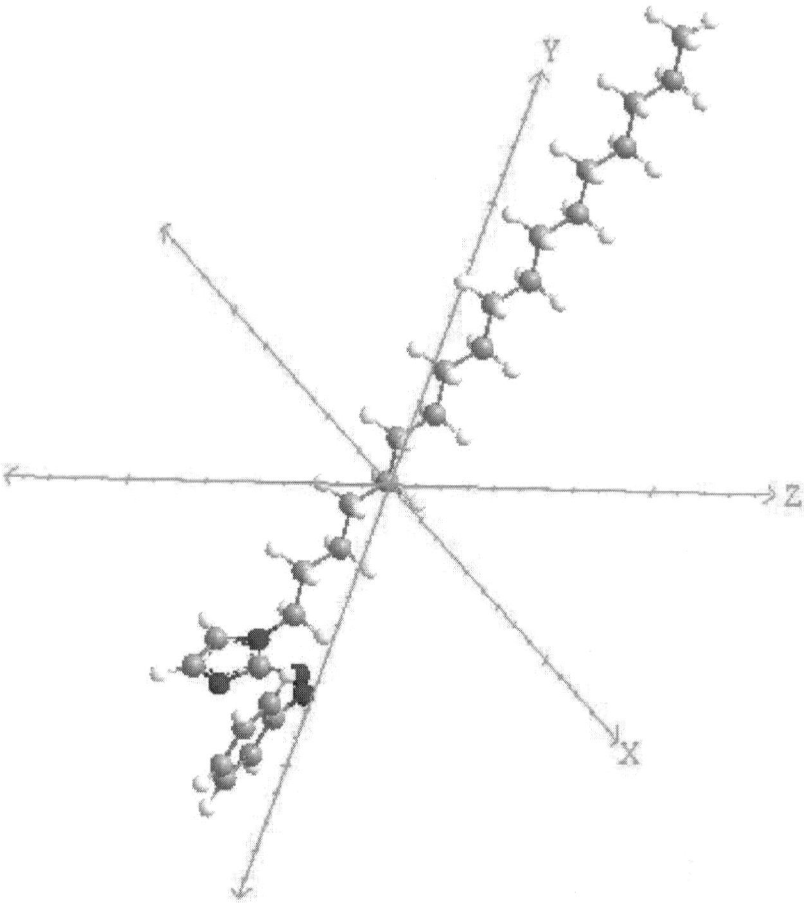

Figure 10. Optimized structure of compound 5a using DFT (B3LYP) level of theory.

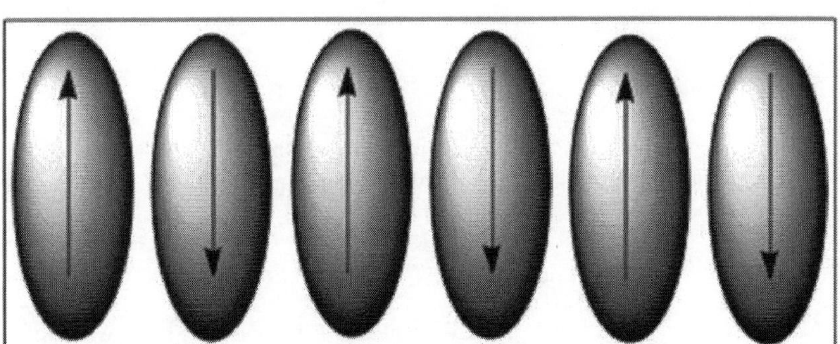

Figure 11. Schematic representation of supramolecular arrangement of dipoles in Nematic phase.

ACKNOWLEDGEMENTS

The authors express their sincere thanks to Department of Science and Technology, New Delhi for financial support. One of the authors (AN) thanks UGC for the award of Fellowship.

REFERENCES

1. R. B. Wesley and B. L. Feringa, "Light Switching of Molecules on Surfaces," Annual Reviews: Physical Chemistry, Vol. 60, 2009, pp. 407-428.

2. N. L. Abbott, C. B. Gorman and G. M. Whitesides, "Active Control of Wetting Using Applied Electrical Potentials and Self-Assembled Monolayers," Langmuir, Vol. 11, No. 1, 1995, pp. 16-18.

3. L. Mu, Y. Liu, S. Y. Cai and J. L. Kong, "A Smart Surface in a Microfluidic Chip for Controlled Protein Separation," Chemistry—A European Journal, Vol. 13, No. 18, 2007, pp. 5113-5120.

4. E. Katz, A. N. Shipway and I. Willner, "Electronic and Optical Transduction of Photoisomerization Processes at Molecularand Biomolecular-Functionalized Surfaces," In: Z. Sekkat and W. Knoll, Eds., Photoreactive Organic Films, Academic, San Diego, 2002, pp. 220-268.

5. M. Ire, "Photochromism: Memories and Switches— Introduction," Chemical Reviews, Vol. 100, No. 5, 2000, pp. 1683-1684.

6. S. Kawata and Y. Kawata, "Three-Dimensional Optical Data Storage Using Photochromic Materials," Chemical Reviews, Vol. 100, No. 5, 2000, pp. 1777-1788.

7. H. Dürr and H. Bouas-Laurent, "Photochromism: Molecules and Systems," Elsevier, Amsterdam, 2003.

8. B. L. Feringa, "Molecular Switches," Wiley-VCH, Weinheim, 2001.

9. J. C. Crano and R. J. Guglielmetti, "Organic Photochromic and Thermochromic Compounds," Kluwer Academic/ Plenum Publishers, New York, 1999.

10. V. El'tsov, "Organic Photochromes," Consultants Bureau, New York, 1990.

11. S. Furumi, H. Fudouzi and T. Sawada, "Dynamic Photoswitching of Micropatterned Lasing in Colloidal Crystals by the Photochromic Reaction," Journal of Materials Chemistry, Vol. 22, No. 40, 2012, pp. 21519-21528.

12. S. Pu, T. Wang, G. Liu, W. Liu and S. Cui, "A New Photoinduced Fluorescent Switch Based on a Photochromic Diarylethene with a Rhodamine Fluorophore," Dyes and Pigments, Vol. 94, No. 3, 2012, pp. 416-422.

13. M. Irie, "Special Issue Photochromism: Memories and Switches," Chemical Reviews, Vol. 100, No. 5, 2000, pp. 1683-1716.

14. S. Kobatake, S. Takami, H. Muto, T. Ishikawa and M. Irie, "Rapid and Reversible Shape Changes of Molecular Crystals on Photoirradiation," Nature, Vol. 446, No. 7137, 2007, pp. 778-781.

15. T. L. Andrew, H. Y. Tsai and R. Menon, "Confining Light to Deep Subwavelength Dimensions to Enable Optical Nanopatterning," Science, Vol. 324, No. 5929, 2009, pp. 917-921.

16. H. Bouas-Laurent and H. Dürr, "Organic Photochromism," Pure and Applied Chemistry, Vol. 73, No. 4, 2001, pp. 639-665.

17. C. V. Yelamaggad, I. Shashikala, U. S. Hiremath, D. S. S. Rao and S. K. Prasad, "Liquid Crystal Dimers Possessing Chiral Rod-Like Anisometric Segments: Synthesis, Characterization and Electro-Optic Behavior," Liquid Crystals, Vol. 34, No. 2, 2007, pp. 153-167.

18. P. M. Maitlis, A. Giroud-Godquin and M. Angecu, "Metallomesogens: Metal Complexes in Organized Fluid Phases," Angewandte Chemie International Edition in English, Vol. 30, No. 4, 1991, pp. 375-402

19. V. Balzani, "Molecular Devices and Machines," WileyVCH, Weinheim, 2008.

20. S. Gerlich, L. Hackermüller, K. Hornberger, A. Stibor, H. Ulbricht, M. Gring, F. Goldfarb, T. Savas, M. Müri, M. Mayor and M. Arndt, "A Kapitza-Dirac-Talbot-Lau Interferometer for Highly Polarizable Molecules," Nature Physics, Vol. 3, No. 10, 2007, pp. 711-715.

21. H. Hartmann and I. Zug, "On the Coupling of Aryldiazonium Salts with N,N-Disubstituted 2-Aminothiophenes and Some of Their Carbocyclic and Heterocyclic Analogues," Journal of the Chemical Society, Perkin Transactions, Vol. 1, No. 24, 2000, pp. 4316-4320.

22. M. M. M. Raposo, A. M. R. C. Sousa, A. M. C. Fonseca and G. Kirsch, "Thienylpyrrole Azo Dyes: Synthesis, Solvatochromic and Electrochemical Properties," Tetrahedron, Vol. 61, No. 34, 2005, pp. 8249-8256.

23. B. A. Trofimov, E. Y. Schmidt, A. I. Mikhaleva, A. M. Vasiltsov, A. B. Zaitsev, N. S. Smolyanina, A. V. Afonin, I. A. Ushakov, K. B. Petrushenko, O. N. K. Kazheva, O. A. Dyachenko, V. V. Smirnov, A. F. Schmidt, M. V. Markova and L. V. Morozova, "2-Arylazo-1-vinylpyrroles: A Novel

Promising Family of Reactive Dyes," European Journal of Organic Chemistry, Vol. 2006, No. 17, 2006, pp. 4021-4033.

24. P. J. Coelho, M. C. R. Castro, A. M. C. Fonseca and M. M. M. Raposo, "Photoswitching in Azo Dyes Bearing Thienylpyrrole and Benzothiazole Heterocyclic Systems," Dyes and Pigments, Vol. 92, No. 1, 2011, pp. 745-748.

25. U. Ray, D. Banerjee, G. Mostafa, T.-H. Lu and C. Sinha, "Copper Coordination Compounds of Chelating ImidazoleAzo-Aryl Ligand. The Molecular Structures of Bis[1- ethyl-2-(p-tolylazo)imidazole]-bis-(azido)copper(II) and Bis[1-methyl-2(phenylazo)imidazole]-bis(thiocyanato)copper(II)," New Journal of Chemistry, Vol. 28, No. 12, 2004, pp. 1437-1442.

26. J. Otsuki, K. Suwa, K. Narutaki, C. Sinha, I. Yoshikawa and K. Araki, "Photochromism of 2-(Phenylazo)imidazoles," The Journal of Physical Chemistry A, Vol. 109, No. 35, 2005, pp. 8064-8069.

27. J. Otsuki, K. Suwa, K. K. Sarker and C. Sinha, "Photoisomerization and Thermal Isomerization of Arylazoimidazoles," The Journal of Physical Chemistry A, Vol. 111, No. 8, 2007, pp. 1403-1409.

28. K. K. Sarker, A. D. Jana, G. Mostafa, J.-S. Wu, T.-H. Lu and C. Sinha, "Iodo Bridged Lead(II) Compounds of Azoimidazoles: Single Crystal X-Ray Structures of [Di-iodo- {1-methyl-2-(p-tolylazo)imidazole}lead(II)]$_n$ and {1- Methyl-3-benzyl-2-(p-tolylazo)imidazolium}$_m$-{tri-iodoplumbate(II)}$_m$," Inorganica Chimica Acta, Vol. 359, No. 13, 2006, pp. 4377-4385.

29. K. K. Sarker, D. Sardar, K. Suwa, J. Otsuki and C. Sinha, "Cadmium(II) Complexes of (Arylazo)imidazoles: Synthesis, Structure, Photochromism, and Density Functional Theory Calculation," Inorganica Chemistry, Vol. 46, No. 20, 2007, pp. 8291-8301.

30. K. K. Sarker, B. G. Chand, K. Suwa, J. Cheng, T.-H. Lu, J. Otsuki and C. Sinha, "Structural Studies and Photochromism of Mercury(II)-Iodo Complexes of Arylazoimidazoles," Inorganica Chemistry, Vol. 46, No. 3, 2007, pp. 670-680.

31. P. Pratihar, T. K. Mondal, A. K. Patra and C. Sinha, "trans-Dichloro-bis-(arylazoimidazole)palladium(II): Synthesis, Structure, Photoisomerization, and DFT Calculation," Inorganica Chemistry, Vol. 48, No. 7, 2009, pp. 2760-2769.

32. D. Demus, J. Goodby, G. W. Gray, H. W. Spiess and V. Vill, "Handbook of Liquid Crystals," Wiley-VCH, Weinheim, 1998.

33. D. Demus, J. Goodby, G. W. Gray, H. W. Spiess and V. Vill, "Dunmur D and Toriyama K. Handbook of Liquid Crystals, Vol 1, Chapter VII," Wiley-VCH, Weinheim, 1998.

34. K. Binnemans, "Ionic Liquid Crystals," Chemical Reviews, Vol. 105, No. 11, 2005, pp. 4148-4204.

35. G. M. Sheldrick, "SHELXS 97, Program for the Solution of Crystal Structure," University of Göttingen, Göttingen, 1990.

36. G. M. Sheldrick, "SHELXL 97, Program for the Refinement of Crystal Structure," University of Göttingen, Göttingen, 1997.

37. L. Spek, "Structure Validation in Chemical Crystallography," Acta Crystallographica Section D, Vol. 65, No. 2, 2009, pp. 148-155.

38. Cambridge Soft Corporation: USA, "Chem3D, Version 10," Cambridge Soft, Cambridge, 2006.

39. M. J. Frisch, "GAUSSIAN 03 (Revision c.02)," Gaussian, Wallingford, 2004.

40. M. M. Enriquez, M. Fuciman, A. M. LaFountain, N. L. Wagner, R. R. Birge and A. H. Frank, "The Intramolecular Charge Transfer State in Carbonyl-Containing Polyenes and Carotenoids," The Journal of Physical Chemistry B, Vol. 114, No. 38, 2010, pp. 12416-12426.

41. P. Bhunia, B. Baruri, U. S. Ray and C. Sinha, "Azoimidazole Complexes of Manganese(II)-Thiocyanate. X-Ray Structures of 2-(p-tolylazo)imidazole and Bis-[1-methyl- 2-(p-tolylazo)imidazole]-di-thiocyanato-manganese(II), " Transition Metal Chemistry, Vol. 31, No. 3, 2006, pp. 310-315.

CHAPTER 4

Past, present, and future of fringe-field switching-liquid crystal display

Dae Hyung Kim[a], Young Jin Lim[a], Da Eun Kim[a], Hongwen Ren[a], Seon Hong Ahn[b] & Seung Hee Lee[a*]

[a] Applied Materials Institute for BIN Convergence, Department of BIN Fusion Technology and Department of Polymer-Nano Science and Technology, Chonbuk National University, Jeonju, Jeonbuk 561-756, Republic of Korea

[b] Samsung Display Co. Ltd., Cheonan, Chungnam 330-300, Republic of Korea

ABSTRACT

The fringe-field switching-liquid crystal display (FFS-LCD) mode recently became a major component of LCDs for high resolution, high-image quality, low power consumption, and touch screen suitability. This review paper discusses how the FFS technology was developed, how it evolved to its present status, and also about the future advances of the FFS mode in terms of performance and technologies.

KEYWORDS

liquid crystal displays, fringe-field switching, high performance

1. ADVENT OF THE FRINGE-FIELD SWITCHING MODE: THE PAST

The first prototype of the thin-film-transistor (TFT) liquid crystal display (LCD) appeared in 1980, and the mass production of the TFT-LCDs started in around the 1990s, based on the twisted nematic (TN) liquid crystal (LC) mode [1–3]. In the beginning, its main application field was displays for desktop monitors and notebook computer screens, which require low power consumption with good image quality at a normal direction. As the TN mode is operated by a low voltage of less than 5 V with good image quality at a normal direction and with a reasonable response time, it was suitable for personal portable displays. In the TN mode, in which the LC is twisted by 90 ° from the top to the bottom substrates, the average LC director tilts upward along only one azimuthal direction so that asymmetry in on-state luminance occurs, resulting in a very narrow viewing angle. There were so many attempts to replace the narrow-viewing-angle TN mode with a wide-viewing-angle mode before the mid-1990s, but as most of them were non-practical, commercialization was not achieved. Nevertheless, as the LCD size became larger than 15″ in around the year 1995, the improvement of the viewing angle of LCDs became a prerequisite. Hitachi Co. reported a very-wide-viewing-angle LCD, the so-called in-plane switching (IPS) TFT-LCD, in 1995 [4–6], and Fujitsu Co. reported a practical multi-domain vertical alignment (MVA) TFT-LCD in 1997 [7,8]. Both became representatives of the wide-viewing-angle LC modes.

Although both the IPS and MVA modes were mainly used by several companies for both the monitor and TV applications at that time, these authors (engineers in the LCD division of Hyundai Electronics Co., now called "HYDIS Co.") believed that both modes have intrinsic problems, such as low transmittance and high operating voltages, which are certainly not suitable for portable display application. Especially, when these authors fabricated a 12.1″ IPS TFT-LCD, it was realized that the device showed an excellent viewing angle but its transmittance was very low with a relatively high operating voltage, such that the conclusion that both modes were not suitable for portable displays was arrived at. Therefore, all kinds of LC modes were tested, along with various electrode structures, to find an optimal solution to the problem of the IPS mode. The fringe-field switching (FFS) idea was then conceived, and its first patent was filed in 1996 [9]. Figure 1 compares the electrode structures of the two modes. The key concept of the IPS mode is the in-plane field driving of homogenously aligned LC molecules. To reorient the LC director by in-plane field, the distance (l) between the electrodes should be larger than the electrode width (w) and cell gap (d). These approaches have two demerits. First, the LC director above the electrodes cannot rotate enough, and as such, transmittance loss above the electrodes cannot be avoided. Second, the operating voltage (V_{op}) of the IPS device is proportional to l/d; that is, it has a larger l, which will give rise to a higher transmittance, but its V_{op} will also be higher. Therefore, achieving IPS-LCDs with high transmittance and low V_{op} at the same time is not possible intrinsically. In the FFS mode, such an intrinsic problem is overcome. By positioning the pixel and common electrodes very close to each other under the electrode conditions (i.e., l/w and l/d<1) or by discarding the conventional l concept such that only the distance (l') between the pixel (common) electrodes exists (l=0 μm), a strong in-plane field that can rotate the LC director will exist between the pixel and common electrodes, and thus, the LC director just above the electrode surface will rotate enough to be able to rotate the LC director above the center of the electrodes, where there is no in-plane field at all. In this way, the LC director above the whole electrode surfaces will rotate, giving rise to a high transmittance in all positions. Therefore, it can be concluded that the LC director in the FFS mode is switched by two mechanisms: first, by the dielectric torque between the LC director and the field, and then by the elastic torque between the LC director at the edge of the electrodes and the LC director at the center of the electrodes [10]. In addition, the field intensity of the FFS mode just above the electrode surface is very strong compared to that of the IPS mode. For example, when V_{op} is 5 V, with l=5 μm, the maximum field intensity felt by the LC is 1 V/ μm in the IPS mode whereas it is 10 V/ μm (or 5 V/ μm when l exists) in the FFS mode because the passivation layer between the pixel and common electrodes is about 0.5 μm (or 1 μm when l exists). In other words, the V_{op} of the FFS mode is much lower than that of the IPS mode. Therefore, the FFS mode can realize a high transmittance and a low V_{op} simultaneously while still exhibiting a wide viewing angle owing to the in-plane rotation of the LC director, which opened a possibility for the FFS mode to be applicable to all kinds of high-performance displays, whether portable or non-portable. Another great advantage of the FFS mode is its self-forming storage capacitor (C_{st}), in which the C_{st} is formed in a light-transmitted area without sacrificing any aperture ratio while all the conventional LC modes need an area for C_{st}, which sacrifices an aperture ratio. This is another main reason that the FFS mode is a suitable one for high-resolution LCDs. Although it seems that the FFS mode very clearly reveals high electro-optic performances, it has one demerit: it requires two transparent electrodes, unlike the other LC modes TN, IPS, and MVA, which require only one transparent electrode in the TFT substrate.

Hereafter, the main milestone of the FFS mode will be discussed.

1.1. First development stage of the FFS mode

The first feasibility of the FFS mode was tested using the conventional 12.1″ TN TFT-LCD by changing a few masks for the FFS TFT-LCD. One active area was divided into four parts: (1) the 1st ITO with a plane shape; and (2–4) the 1st ITOs with slit shapes. The first prototype was working fine while showing a much higher transmittance than the IPS-LCD, as presented in Figure 2. The transmittance of the FFS mode with l=0 μm or 1 μm showed the highest transmittance when an LC with negative dielectric anisotropy (−LC) was used. When the feasibility of the FFS LCD was confirmed in 1998, a 15.0″ XGA TFT-LCD using an LC with negative dielectric anisotropy was developed for monitor displays. Interestingly, the device showed the best transmittance among the wide-viewing-angle LCDs at that time, and as such, it was said to be the device that shows a wide viewing angle and a high transmittance at the same time in the authors' talk upon the invitation of Asia Display [11–13].

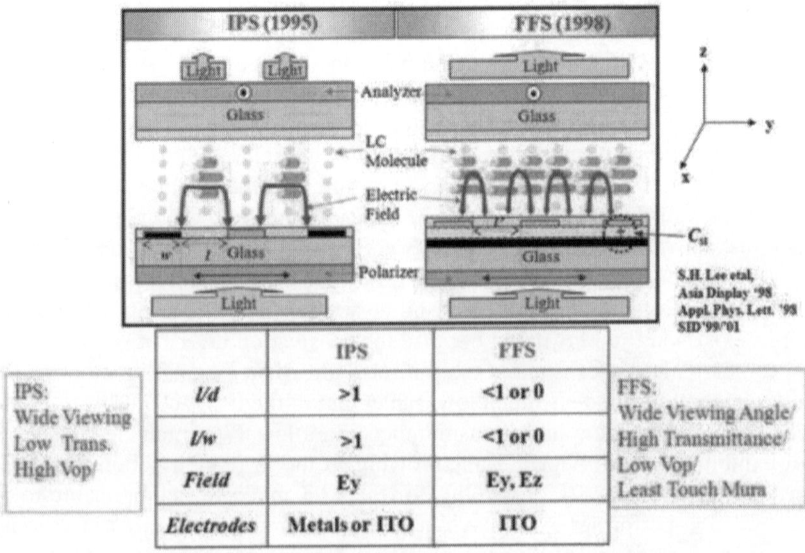

Figure 1. Schematic comparison of the electrode structures of the IPS and FFS modes.

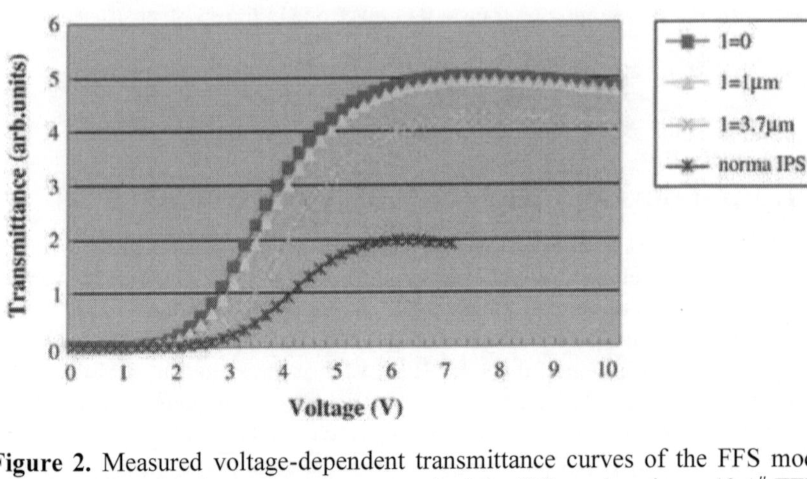

Figure 2. Measured voltage-dependent transmittance curves of the FFS mode with four different electrode structures and of the IPS mode using a 12.1″ TFT-LCD. Here, the IPS and FFS modes use a LC with positive and negative dielectric anisotropies, respectively.

Another unique and interesting performance aspect of the LCD was found. That is, the display almost did not show ripples and pooling mura or pushed marks at all when an external pressure was applied to the panel device with the fingers or a pen [14,15]. In the case of the TN mode, strong ripples can be easily observed when pressure is applied to the panel, and in the case of most VA modes, relatively strong pooling mura can be observed. At that time, Wacom Co., which makes pen-based 3D input displays for animation drawing works in Hollywood movies noticed the strong merit of the FFS displays and commercialized them for display application for the first time. When the company used the FFS mode, the demerits of the conventional display, which requires a thick cover glass with an air gap between the panel and the cover glass, causing image parallax to occur with weight and system cost increases while decreasing the transmittance, were overcome. They thus received major compliments from their customers, as follows: "Tablets with conventional LCDs have a parallax issue and feel like drawings on a glass, but with the FFS Tablet PC, there is no parallax issue, and we feel like we are drawing on a piece of paper." In fact, this pen-based display indicates that the FFS-LCD is suitable for the touch screen display, which is a required function in all portable displays. This fact was already recognized in 1998.

1.2. Second development stage of the FFS mode

Since the mid-1990s, the size of the TFT-LCDs for monitor use has become increasingly bigger than 17″, and the monitor has also come to play the role of a TV at the same time. The first FFS-LCD used a −LC with a rather high cost and a much larger rotational viscosity than that of the LC with positive dielectric anisotropy (+LC). Moreover, the magnitude of its dielectric anisotropy for a commercialized −LC mixture is much lower than that of +LC. Therefore, the FFS-LCD with −LC would show a slower response time and a higher operating voltage than that with +LC. If a +LC with IPS mode concepts will be used, however, the transmittance of the FFS-LCD will no longer be excellent, and a certain level of pooling mura will appear. At that time, there were extensive studies regarding what determines the transmittance of the FFS-LCD with +LC, and it was found that the light efficiency of the LC mode was dependent on many cell parameters, such as the LC's retardation [16], the cell gap [17], the electrode structure [18], and the magnitude of dielectric anisotropy [19]. In addition, it is widely known that the transmittance of the LCD is dependent not only on the LC's light efficiency but also on the pixel aperture ratio. As such, all the panel and cell parameters were optimized to maximize the performance of the FFS-LCD with +LC, and "U-FFS" [20], which exhibits the following properties, was proposed:

- high transmittance;
- high contrast in a wide range;
- minimized color shift;

- fast response; and
- low power consumption.

As presented in Figure 3 for the pixel structure, the pixel electrode has a wedge shape with a symmetry line in the horizontal direction, so that the initial LC orientation is in a horizontal direction, consistent with the field direction between the data line and the pixel electrodes, which can keep an original dark state. Owing to such characteristic, the width of the black matrix on the top substrate can be minimized, resulting in a maximized aperture ratio. The proposed concept becomes the basis of all FFS-LCDs, including the tablet PCs in early 2000, the black-and-white displays for medical uses, and the monitors and LCD-TVs in the mid-2000s.

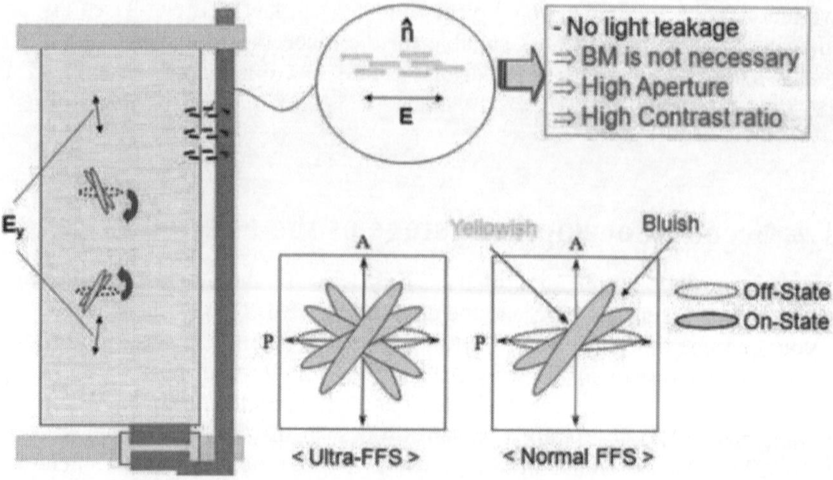

Figure 3. Schematic pixel structure of the U-FFS LCD.

1.3. Third development stage of the FFS mode

In the early 2000s, tablet LCDs with a pen-based system with the TN mode already existed, but the TN mode had a very narrow viewing angle, which was not suitable for the tablet PC because it could be displayed in either the landscape or portrait type depending on the user's preference. In addition, the device must be high-resolution, like XGA in the 12.1" display, so that maximizing the transmittance while keeping a wide-viewing angle and a low operating voltage was very important for high-performance LCDs with a high image quality and low power consumption. As the pixel size became smaller, the unwanted field direction around the corners of the pixel electrode no longer became negligible, and the stability of the LC's reorientation with +LC when an external pressure was applied was not satisfactory enough. A better pixel structure, which solved this problem, and maximized transmittance in relatively high-resolution FFS-LCDs were proposed as A-FFS [21]. In the mid-2000s, Hydis Co. and Seiko Epson tried to apply the FFS mode to super-high-resolution mobile LCDs. Then they proposed a revolutionary pixel structure that maximizes the transmittance in the FFS mode, the so-called “H-FFS” [22]. As clearly shown in Figure 4, the transmittance of the 2.0″ QVGA HFFS-LCD was improved by more than 50% compared to that of the AFFS-LCD, and was even higher than that of the TN-LCD.

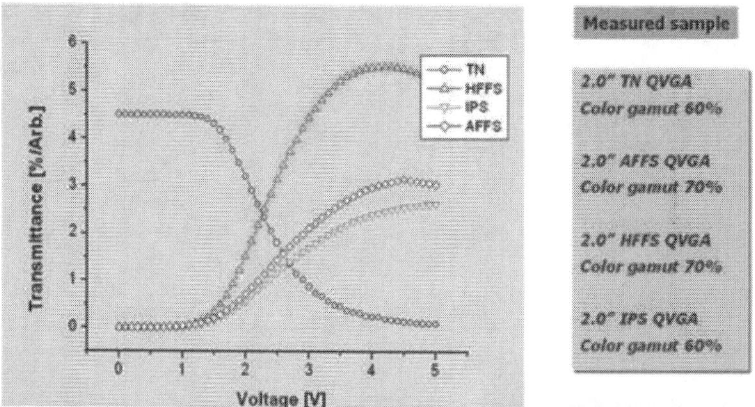

Items	2.03" AFFS	2.03" HFFS
Resolution	QVGA	QVGA
Color Gamut	NTSC 70%	NTSC 70%
Luminance	200 nit	312 nit
Contrast Raito	450 : 1	650 : 1
X-talk (H/V)	1.0% / 4.5 %	1.0%/1.0%
Transmittance (No gain)	4.0 % →	6.2 %

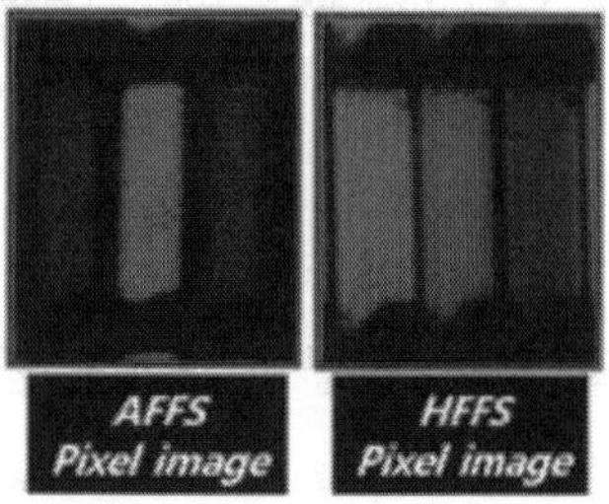

Figure 4. Comparison of the voltage-dependent transmittance curves in high-resolution mobile TFT-LCDs. The HFFS mode shows a much improved transmittance, as also shown in the microscopic image of a pixel [21] and see also the website of HYDIS Co.

The key feature of the structure is that the common electrode exists above the pixel electrode, unlike that in the conventional AFFS mode, and covers the data line to shield off any noise field coming from the data line, so that the transmittance near the data line is maximized by removing the LC's disclination lines and minimizing the width of the black matrix on the top substrates, as indicated in Figure 5. The HFFS structure has become the foundation of all the present high-resolution FFS (IPS) LCDs, from small-to-medium-sized displays.

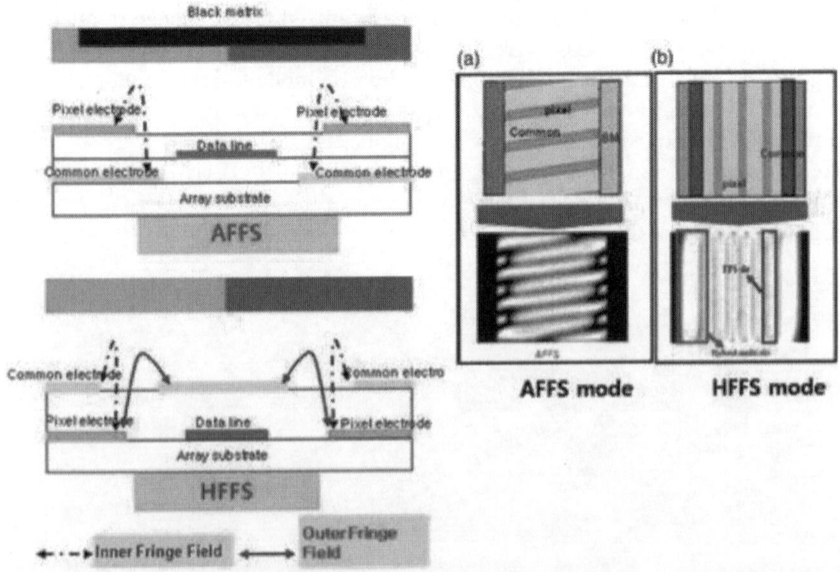

Figure 5. Schematic comparison of the pixel structures and simulated transmittances of the AFFS and HFFS modes [21].

2. HISTORICAL MOVEMENT OF THE FFS MODE

2.1. First movement

The FFS mode overcomes the long-standing problems of the IPS mode. Nevertheless, the FFS mode was not adopted by other major companies until 2004. Instead, Hitachi Co. continuously reported the improved performance of the IPS TFT-LCD and S-IPS in 1998 and of the AS-IPS in 2002. The AS-IPS used an organic resin to improve the transmittance, which was not an easy process at that time. In addition, the resin was opened between the electrodes and can become a source of ions, causing image sticking. If the FFS mode is adopted without using resin, however, its transmittance will be higher than that of the AS-IPS, with a lower operating voltage. Consequently, Hitachi Co. adopted a new IPS, with the same concept as that of the FFS mode (they named it “IPS-Pro”), in 2004 [23], which surprised the world's display industries. The FFS mode was then noticed more by other companies.

2.2. Second movement

The major expert company for small LCDs in the early 2000s was Seiko-Epson Co. The company also adopted the HFFS mode for their super-high-resolution mobile LCDs in 2006, which marked FFS's first application to mobile LCDs with high-resolution LCDs.

2.3. Third movement

Until the first and second historical movements, the FFS mode was still not in the main LC mode because the biggest LCD companies, such as LGD, Samsung, Sharp, and AUO, did not use the FFS mode, although LGD used the FFS mode for its mobile LCDs. In 2010, revolutionary products called "iPhone4 Retina Display" and "iPad" were released by Apple Co. At that time, the required resolution of the retina display was the highest, 326 ppi, while keeping low power consumption and high image quality. On the other hand, the required electro-optic performances of the iPad product were as follows: 10-h use without recharging the battery, wide-viewing angle, and touch screen display. According to the analyses conducted by Apple Co., the FFS-LCD best met their requirements, and as such, the LCD panel maker LGD started its mass production with the FFS mode for the iPhone and iPad. Since then, the FFS-LCD has become the symbol of the high-resolution and high-performance TFT-LCDs.

3. POPULARITY OF THE FFS LC MODE: THE PRESENT

When Apple Co. used the FFS mode for its products with a high resolution, high image quality, low power consumption, and touch screen displays, the merits of the FFS mode were recognized by many other companies. Trials were conducted to meet such requirements using the conventional IPS and MVA modes, but the performances of such modes could not meet the level of the FFS mode.

Since 2010, most of the high-resolution and high-image-quality LCDs for mobile phones and tablet PCs have adopted the FFS mode. On the other hand, TN and WV-TN were the main LC modes for the notebook and monitor displays. As the display resolution became increasingly higher for high image quality, however, the transmittance of the TN mode was not better than that of the FFS mode, and the viewing angle characteristics and touch screen suitability of the TN mode were inferior to those of the FFS mode. Since recently, the most important keyword in displays has become "high resolution" because the display image quality in the normal direction purely depends on the resolution, irrespective of the display mode or type. From this viewpoint, the FFS mode is the most suitable mode because it exhibits the highest transmittance among all the LC modes, especially in LCDs with a high resolution. On the other hand, the LCDs are competing with the emissive OLEDs, whose image quality does not have many viewing angle issues. Consequently, using only the TN mode cannot realize high-performance LCDs, as above-mentioned, and as such, the TN mode is currently being replaced by the FFS mode even in notebooks and monitors. When the HFFS mode is used for large super-high-resolution FFS-LCDs, however, the electrical load becomes too high, which makes driving very difficult. As a result, several companies, including Panasonic LCD Co., proposed a new electrode structure employing an organic resin above the data line, and the plane shapes of the common electrodes covered all the signal lines to shield any noise electric field coming from the TFT, as shown in Figure 6 [24,25]. In Figure 6, the insulator between the data line and the common electrode is not indicated as an organic resin, but for large (over 10″) and super-high-resolution TFT-LCDs, organic resin is suitable for the over-1-μ m layer. Panasonic LCD Co. claimed that the high-transmittance FFS-LCD can be realized even if the display resolution increases from the full-HD to the 4K2K proposed structure, as shown in Figure 7. Even if the resolution is increased fourfold, the width of the black matrix on the top substrate can be reduced with the proposed structure, so that the aperture ratio is maintained at 70%, which is about the same value of the full-HD structure. In the conventional LCD concepts and approaches, this is not imaginable; that is, only the FFS mode, which has an optimized pixel structure, can open a new paradigm: that increasing the resolution of LCDs fourfold will not decrease the aperture ratio. For LCD-TVs, Japan/Panasonic LCD Co. and BOE Co. have adopted the FFS mode for their LCD-TVs. In addition, BOE Co. recently reported the largest 110″ UD TFT-LCD with the FFS mode, with the following panel specifications: $\text{brightness} = 1000$nits; contrast $\text{ratio} = 1200:1$; response time=6 ms (G to G); and frame $\text{rate} = 120$ Hz [26]. Surprisingly, a-Si for TFT and aluminum (Al) metal for signal lines are used even with the 110″ LCD display.

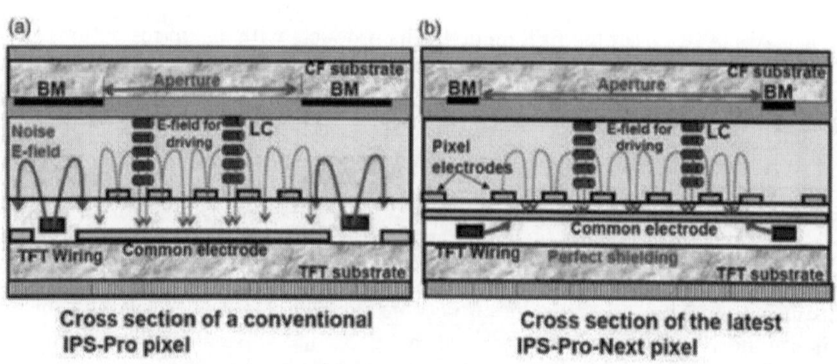

Figure 6. Comparison of the pixel structures of the IPS-Pro and IPS-Pro Next pixels [23,24].

To conclude, all the current high-resolution and high-performance LCDs, from the small to the medium-sized ones (less than 30″), are mainly using the FFS mode, and surprisingly, most of the major LCD companies at present are using the FFS mode for such applications, although they use different names for it, such as IPS-Pro, Photo Fine Vistarich, AH-IPS, PLS, and ADS.

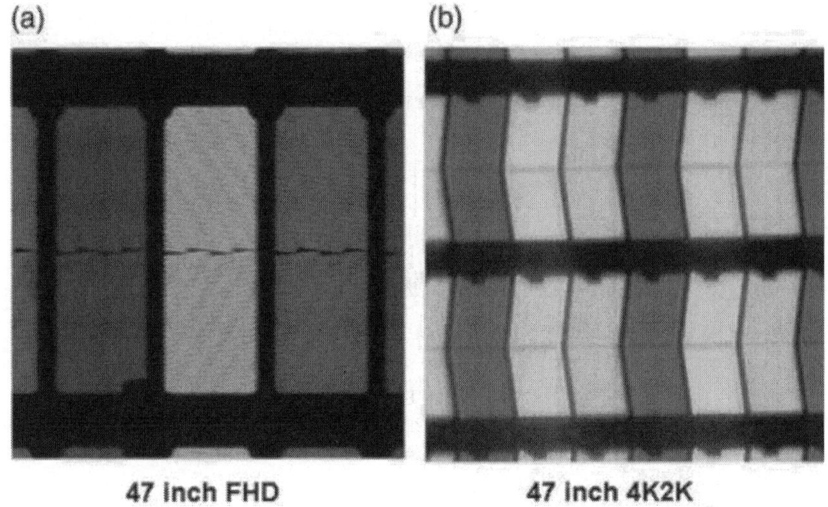

Figure 7. Comparison of the pixels of the 47″ IPS-LCDs [24].

4. TECHNICAL ISSUE OF THE PRESENT FFS-LCD

4.1. LC

At present, most of the FFS LCDs use a +LC because it offers a lower operating voltage, a faster response time, and a lower cost than a −LC. The transmittance, however, becomes more critical as the resolution of LCDs becomes higher — i.e., over 500 ppi for the mobile and retina displays of medium-sized displays. To maintain the high transmittance of the FFS-LCD even in super-high-resolution LCDs, a −LC is favored, even though its response time will be slower and its operating voltage will slightly increase. Therefore, further optimization of the cell parameters and LC materials is absolutely required for the high-transmittance and fast-response-time FFS-LCD with −LC [27].

4.2. Non-contact LC alignment

The FFS-LCD requires a homogenous LC alignment with a zero pretilt angle. Most of the FFS-LCDs are made with a rubbing process so that rubbing-related defects always exist, resulting in a decrease in the contrast ratio of the LCD. Moreover, the rubbing generates a 1–2 ° surface pretilt angle. Therefore, this rubbing process should be replaced by a non-contact LC alignment process. Several types of photoalignment layers are currently being tested, and Panasonic

LCD Co. reported photoaligned IPS-LCDs for TV use, which showed a 1700:1 contrast ratio [28,29].

4.3. Further optimization of electro-optic performances

The FFS mode shows some unusual electro-optic behavior so that the DC-driven voltage-dependent transmittance curve between the positive and negative frames does not perfectly match, unlike in the other conventional LC modes. In addition, low-frequency driving, such as at 5 Hz, generates flickering. Further extensive studies are thus required to fully understand this behavior. In addition, especially when a −LC is used, a thin cell gap (less than 3 μm) and fine electrode patterning (close to 2 μm) are strongly favored to achieve fast-response-time and high-transmittance FFS-LCDs [27,30].

4.4. Charging issue in large LCD-TVs

The FFS mode has a great advantage in the pixel aperture ratio because of its self-formed storage capacitor, as mentioned earlier. The C_{st} might become too big, however, when the pixel size is quite large, as in the LCD-TV, and thus, a charging issue appears. Although this kind of intrinsic difficulty exists, BOE Co. already proved the perfect driving of 110″ UD FFS-LCD using an a-Si TFT, which implies that the intrinsic problem can be overcome with the optimization of the panel and driving. To solve the charging issue, an oxide TFT with high mobility needs to be adopted, and the pixel structure needs to be optimized, especially in large FFS-LCDs.

5. VISION OF THE FFS LC MODE: THE FUTURE

The FFS-LCD is continuously evolving to exhibit better performance as well as low cost, targeting better performance in electro-optics and lower cost over the emissive OLEDs. Looking back on the major historical movement of the FFS-LCD since it was first reported to the public in 1998, every six years, a great event occurred, such as that the Hitachi LCD-TVs adopted the FFS in 2004 and Apple iPad adopted the FFS mode in 2010. It is hoped that eco-friendly green FFS LCD-TVs with photoalignment, oxide TFTs, and a revolutionary LC mixture will become the main trend even in the LCD-TVs that will be produced by many companies in 2016.

ACKNOWLEDGEMENT

This work was supported by the National Research Foundation (NRF) of Korea and the Korea-China Joint Research Program under grant no. 2012-0004814.

REFERENCES

1. M. Schadt and W. Helfrich, Appl. Phys. Lett. 18, 127 (1971).
2. S.H. Lee, S.S. Bhattacharyya, H.S. Jin and K.U. Jeong, J. Mater. Chem. 22, 11893 (2012). doi: 10.1039/c2jm30635b
3. H.S. Kwok, S. Naemura and H.L. Ong, Progress in Liquid Crystal Science and Technology: in Honor of Shunsuke Kobayashi's 80th Birthday (World Scientific'13, Singapore) Series on Liquid Crystal 4 (2013).
4. M. Oh-e, M. Ohta, S. Aratani and K. Kondo, in *Proceedings of the 15th International Display Research Conference (Asia Display)* (1995), p. 577.
5. M. Ohta, M. Oh-e and K. Kondo, in *Proceedings of the 15th International Display Research Conference (Asia Display)* (1995), p. 707.
6. M. Oh-e and K. Kondo, Appl. Phys. Lett. 67, 3895 (1995).
7. Y. Koike, S. Kataoka, T. Sasaki, H. Tsuda, A. Takeda and K. Ohmuro, in *Proceedings of the 4th International Display Workshops* (1997), p. 159.
8. Takeda, S. Kataoka, T. Sasaki, H. Chida, H. Tsuda, K. Ohmuro, T. Sasabayashi, Y. Koike and K. Okamoto, SID Symp. Digest Tech. Pap. 29, 1077 (1998).
9. S.L. Lee and S.H. Lee, KR Patent No. 19,960,059,509 (2000).
10. S.H. Jung, H.Y. Kim, M.-H. Lee, J.M. Rhee and S.H. Lee, Liq. Cryst. 32, 267 (2005).
11. S.H. Lee, S.L. Lee and H.Y. Kim, in *Proceedings of the 18th International Display Research Conference (Asia Display)* (1998), p. 371.
12. S.H. Lee, S.L. Lee and H.Y. Kim, Appl. Phys. Lett. 73, 2881 (1998).
13. S.H. Lee, S.L. Lee, H.Y. Kim and T.Y. Eom, SID Symp. Digest Tech. Pap. 30, 202 (1999).
14. S.H. Lee, H.Y. Kim, S.M. Lee, S.H. Hong, J.M. Kim, J.W. Koh, J.Y. Lee and H.S. Park, J. Soc. Inf. Display 10 (2), 117 (2002).

15. J.D. Noh, H.Y. Kim, J.M. Kim, J.W. Koh, H.S. Park and S.H. Lee, in *Proceedings of the 2nd International Display Manufacturing Conference* (2002), p. 447.

16. I.H. Yu, I.S. Song, J.Y. Lee and S.H. Lee, J. Phys. D: Appl. Phys. 39, 2367 (2006).

17. H.Y. Kim, S.H. Hong, J.M. Rhee and S.H. Lee, Liq. Cryst. 30, 1285 (2003).

18. I.H. Yu, I.S. Song, J.Y. Lee and S.H. Lee, J. Phys. D: Appl. Phys. 39, 2367 (2006).

19. J.W. Ryu, J.Y. Lee, H.Y. Kim, J.W. Park, G.D. Lee and S.H. Lee, Liq. Cryst. 35, 407 (2008). S.H. Lee, S.M. Lee, H.Y. Kim, J.M. Kim, S.H. Hong, Y.H. Jeong, C.H. Park, Y.J. Choi, J.Y. Lee, J.W. Koh and H.S. Park, SID Symp. Digest Tech. Pap. 32, 484 (2001).

20. Y. Nakayoshi, N. Kurahashi, J. Tanno, E. Nishimura, K. Ogawa and M. Suzuki, SID Symp. Digest Tech. Pap. 34, 1100 (2003).

21. D.H. Lim, H.Y. Lee, J.P. Kim, K.J. Lee, H.J. Yun, K.Y. Han, Y.H. Jeong, in *Proceedings of the 13th International Display Workshops* (2006), p. 807.

22. K. Ono, I. Mori, Y. Tomioka and Y. Satou, in *Proceedings of the 15th International Display Workshops* (2004), p. 295.

23. K. Ono and I. Hiyama, in *Proceedings of the 19th International Display Workshops in Conjunction with Asia Display* (2012), p. 933.

24. K. Ono and H. Matsukawa, in *Proceedings of the 20th International Display Workshops* (2013), p. 24.

25. Y.P. Liao , in *Proceedings of the 13th International Meeting on Information Display* (2013), p. 11.

26. H.J. Yun, M.H. Jo, I.W. Jang, S.H. Lee, S.H. Ahn and H.J. Hur, Liq. Cryst. 39, 1141 (2012). H.J. Ahn, C. Lim, D. Kim, J. Lee, H. Park, S. Lee, J. Woo, W. Shin and M. Jun, SID Symp. Digest Tech. Pap. 44, 1432 (2012).

27. H. Matsukawa, Y. Shimano, M. Suefuji, Y. Umeda and K. Mimura, in *Proceedings of the 19th International Display Workshops in conjunction with Asia Display* (2012), p. 1581.

28. S.H. Lee, SID Seminar note, M4 (2013).

CHAPTER 5

Technical evolution of liquid crystal displays

Kyeong-Hyeon Kim[1] and Jang-Kun Song[1]

[1]LCD Division, Samsung Electronics Co. Ltd, 200 Myeongami-Ri, Tangjeong-Myeon, Asan City, Chungcheongnam-Do 336-841, Korea

ABSTRACT

Liquid crystal displays (LCDs) have evolved rapidly as a result of fierce competition among the various LCD technologies, and now occupy the largest proportion of the entire display market. The evolution of LCDs continues, with new technologies and new materials in development to replace current devices. This review summarizes the key technologies used in commercially successful LCD products, focusing on the requirements for high-end displays and the benefits of the in-plane switching and multi-domain vertical alignment modes. As in past advances, the development of new materials will play an important role in the continued technical evolution of LCDs.

In 1888, more than a century ago, the Austrian botanist Friedrich Reinitzer discovered an intermediate state of matter between isotropic liquid and lattice-structured crystal.[1,2,3] This state, later termed liquid crystal (LC), remained merely a novelty for many years until suitable applications for the technology began to become apparent. From the late 1960s, following the realization that various display applications were possible, the popularity of liquid crystal as a research field expanded rapidly, attracting the interest of many scientists who made fundamental contributions to the technologies that have led to the spectacular success of liquid crystal displays (LCDs).[3,4,5,6,7,8,9,10,11,12,13] The development of the twisted nematic (TN)[4] and super-twisted nematic (STN)[5] cell configurations, along with the necessary manufacturing technologies,[6,7,8,9] resulted in the birth of the LCD industry in the late 1970s and early 1980s. The subsequent development of thin-film transistor (TFT) technology provided a further boost to the industry in the late 1980s.[14,15] The boom was significantly supported by the advancement of material technologies such as the highly reliable manufacturing of liquid crystal base materials, and the development of polymers for the alignment layer, color filter materials, and sheet-type polarizers formed from the poly(vinyl alcohol)-iodine complex.

In the 1990s, LCDs occupied an important position in the display market primarily through notebook applications, although the cathode ray tube (CRT) continued to dominate the desktop monitor and television markets. The plasma display panel (PDP), which appeared in the commercial market in 1997, pioneered the large-area flat panel display market. It was widely predicted at the time that PDPs would maintain the lead in the large-area display market and that LCDs in small-display information technology (IT) applications would soon be replaced by new technologies such as displays based on organic light-emitting diodes (OLEDs).[16,17] Such predictions were based on the belief that the LCD technology suffered from serious limitations in terms of both off-axis image quality and moving picture quality, and also that it had proved difficult to obtain reasonable manufacturing yields of large-area TFT panels. However, these predictions, as we now know, turned out to be incorrect. The application of LCDs has expanded rapidly in recent years from purely information technology (IT) applications to the television and mobile display markets, and now even to very-large-area digital information displays (DIDs).

Interestingly, while LCDs penetrated the traditional territories of other display technologies including CRTs and PDPs, attracting inevitable and heavy competition from these technologies, the most critical competitors were in fact alternative LCD technologies. A range of LCD modes and technologies are now applied in the same areas of application and are compared directly with one another in the market. In applications for which LCD technology always held the monopoly, such as small display devices, LCDs avoided direct competition with other displays technologies. However, the competition among the various LCD technologies has been unavoidable and persistent, and it is through this fierce competition that LCD technologies have evolved so rapidly. This strongly market-driven evolution is one of the main reasons that the early predictions regarding the future of LCDs were proved incorrect. Here, we briefly review the history of technical evolution in LCD technology from the point of view of the performance required for high-end displays.

1. WIDE VIEWING ANGLE PERFORMANCE

1.1 Twisted nematic mode

The TN mode is currently one of the most widely used modes for LCD panels.[4] Figure 1 shows a schematic of a TN-mode TFT LCD. Many of the components comprising the panel are common to all LCDs. The LC layer is injected between a TFT substrate and a color filter (CF) substrate. The TFT circuits (data and gate bus lines), pixel electrodes and TFT channels are embedded in the TFT substrate, and the CF substrate consists of sequentially stacked black matrix (BM), color filter (CF) and transparent common-electrode layers. Two orthogonally aligned polarizers are attached to the outer faces of the panel. The LC molecules are aligned by surface anchoring to alignment layers formed on the inner surfaces of the two substrates by rubbing in one direction, which results in a preferred alignment of the LC molecules. The rubbing directions for the two substrates are at right angles in the TN mode, resulting in an LC layer with a 90° twisted structure in the absence of an electric field. The twisted LC structure rotates the polarization of light passing through the cell, aligning the polarization with that of the front polarizer and allowing light to pass (bright state). On increasing the electric field applied between the pixel and common electrodes, the LCs align with the electric field, reducing the birefringence (polarization rotation) of the LC layer and resulting in a gray or dark pixel state. Thus, LCDs based on the TN structure exploit the birefringence effect of the LC layer directly to control the polarization of light and the transmittance of the panel. However, as the apparent birefringence of the LC layer varies with the viewing direction, image distortion occurs in off-axis observation. This degradation of off-axis image quality is generally considered to be the weakest point of LCD technology, and has been addressed by the development of new LC technologies. Nevertheless, the TN mode is simple and offers good brightness and response time performance.[4]Hence, the TN mode has been the most widely commercialized LCD mode.

For high-performance applications, a number of new LCD modes providing an improved range of optimum viewing angle have been introduced, including the multi-domain TN mode, surface-stabilized ferroelectric liquid crystal (SSFLC), antiferroelectric liquid crystal (AFLC) and the optically compensated birefringence (OCB) mode.[10,11,12,13,18,19,20,21] Most of these technologies, however, have not been successfully commercialized due either to difficulties in mass production or to drawbacks other than poor viewing angle. The two technologies that have achieved commercial success are the in-plane switching (IPS) mode and the multi-domain vertical alignment (VA) mode.

1.2 In-plane switching mode

In the IPS mode (Figure 2), which was first suggested in 1992, uniformly aligned LC molecules with positive dielectric anisotropy are rotated azimuthally on a glass plate by an electric field aligned parallel to the plate.[13,21,22,23] The electric field for azimuthal switching is applied via interlaced electrodes formed on one of substrates. The fringe field switching (FFS) mode (also shown in Figure 2) is an advanced form of the IPS technology [24]. The transmittance of light in the normal direction of the LC cell, which contains a uniformly aligned LC layer, can be expressed simply by

$$T = T_0 \sin^2[2\theta(V)]\sin^2[\pi d\Delta n(V)/\lambda] \quad (1)$$

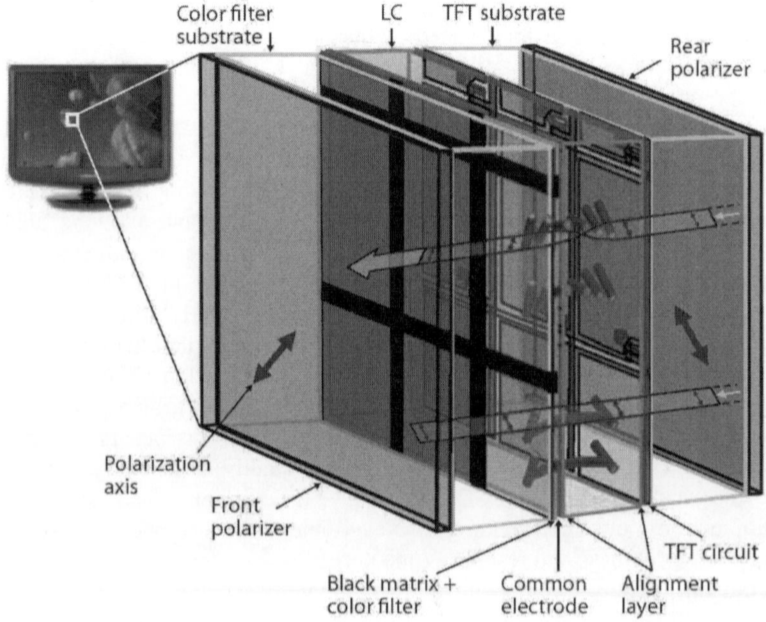

Figure 1. Schematic diagram showing the construction of a TN-mode TFT LCD panel. The twisted LC structure rotates the polarization of incident light (upper pixel) without an applied electric field, aligning the polarization with the front polarizer (bright state). When an electric field is applied (lower pixel), the LC layer becomes homeotropically aligned and the polarization of light is no longer aligned with the front polarizer (dark state).

Here, θ is the angle between the optical axis of the LC layer and the direction of the polarizer, V is voltage, d is the cell thickness or the length of the light path, Δn is the optical anisotropy and λ is the wavelength of light. In the IPS mode, the angle θ, and hence the light-transmittance of the cell, is controlled by the applied electric field. The alignment layer is rubbed parallel to the axis of the polarizer to align the LC molecules in the dark state. The angle θ varies with the strength of the applied electric field, and non-zero transmittance (*i.e.* gray levels) is obtained according to equation (1). Hence, gray levels are determined by the first sine term in equation (1), dependent only on θ. When viewed from an off-axis horizontal angle, θ should be replaced by the effective angle θ' projected on a plane perpendicular to the ray direction in the LC layer. The second sine term containing Δn and λ remains largely constant with respect to the electric field. The first sine term is not sensitive to the viewing direction, since the molecules are oriented parallel to the substrate. Overall, the IPS mode is therefore less sensitive to viewing angle than the TN mode, making the IPS mode a wide-viewing-angle mode.[25]

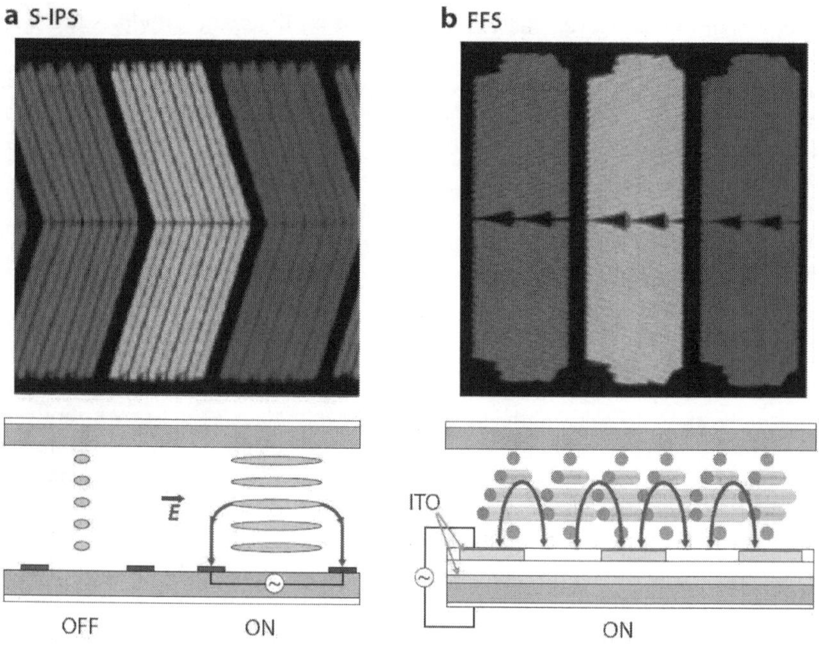

Figure 2. Microscope images (upper) and schematic diagrams (lower) of pixels with the (a) S-IPS and (b) FFS structure.

However, the second sine term in equation (1) is somewhat dependent on the viewing angle, and results in a yellowish image in one diagonal direction and a bluish image in the other diagonal direction. To compensate for this color shift, a chevron-shaped pixel structure with an obtuse angle is used (Figure 2). This two-domain structure is called the super IPS (S-IPS) mode.[26]

FFS panels have two transparent electrode layers, one of which contains a fine grid that induces electric fringe fields (see Figure 2). These fringe fields drive all the LC molecules in the pixel area, including LC on the transparent electrode. Compared to the IPS mode, which uses opaque metal electrodes, the FFS offers higher transmittance due to the high aperture ratio and the fringe field effect, although the preparation of the panels requires at least one additional photo-lithographic process.[24]

1.3 Vertical alignment mode

In the VA mode,[10,12] LC molecules with negative dielectric anisotropy are aligned vertically on the substrate in the absence of an electric field, corresponding to the dark state. Under an electric field, the molecules tilt over, resulting in the bright state. The transmittance of the VA cell is also expressed by equation (1). However, the term dependent on the applied field is not θ but the effective birefringence, $\Delta n(V)$, in the second sine term. $\Delta n(V)$ is sensitive to the viewing direction in the VA mode, since the molecules are inclined on the glass substrate under intermediate electric fields. Hence, the VA mode was not a wide-viewing-angle mode when it was first introduced in 1971.[10] Due to a lack of availability of reliable LC and alignment materials, as well as the poor quality of the viewing angle, the VA mode was not used in commercial products for a long time. These limitations were finally overcome in the late 1990s, and the VA mode became one of the major commercial wide-viewing-angle modes. The key technologies leading to these improvements are a multi-domain structure and a suitable compensation film.[27,28]

It is a rather simple engineering task to control the tilt direction of molecules and to form multiple domains in the VA mode. Since molecules in the VA mode are aligned vertically in the dark state, the molecules can be tilted over in any direction by even a weak asymmetric force. The tilt direction of LC molecules can thus be controlled using simple structures. In the multi-domain VA (MVA) mode, protrusions are added to the electrodes to force the molecules to tilt from the surface, as shown in Figure 3.[27] The patterned VA (PVA) mode uses open slits on the pixel electrodes to induce fringe fields on the edge of the slits, which causes the molecules to tilt over in the direction perpendicular to the slit (Figure 3).[28] Thus, four-domain structures can be achieved relatively easily by optimizing the design of either the protrusion (in the MVA mode) or the indium tin oxide (ITO) slit (in the PVA mode). The four sub-domains compensate for one another, enhancing image quality under horizontal off-axis viewing. However, although the four-domain structure improves the image quality significantly, the performance is still insufficient for high-performance LCDs. In particular, color wash-out in off-axis observation needed to be addressed. The super PVA (S-PVA) mode, a mode with eight sub-domains, was suggested in response to this need (Figure 3).[29] Each pixel in an S-PVA panel is composed of two subpixels, each with four domains. Different levels of electric fields are applied to each subpixel, resulting in a different tilt angle of the LC molecules. Thus, each pixel has eight effective domains. The S-PVA mode has much improved image quality in off-axis observation. An advanced MVA technology with eight domains per pixel has also been developed and is now used in MVA applications.

The eight-domain, two-subpixel structure for S-PVA and advanced MVA panels is implemented in a number of ways. Two-TFT technology [30] is one example in which each subpixel has a dedicated TFT that is linked to the gate or data bus lines. Optimized image quality is thus obtained by controlling each subpixel independently. This is the simplest method for constructing the eight-domain format, but requires the number of data or gate bus lines to be doubled. To avoid this problem, other technologies such as charge sharing, capacitive-coupling, advanced capacitive-coupling and common voltage swing methods have been developed.[30,31,32] Although none of these methods requires the number of gate or data bus lines to be increased, the methods offer less control over the electric voltage applied to each subpixel. Thus, two-TFT technology has the greatest advantage in terms of off-axis image quality, even though it has the apparent disadvantage of doubling the number of gate or data bus lines.

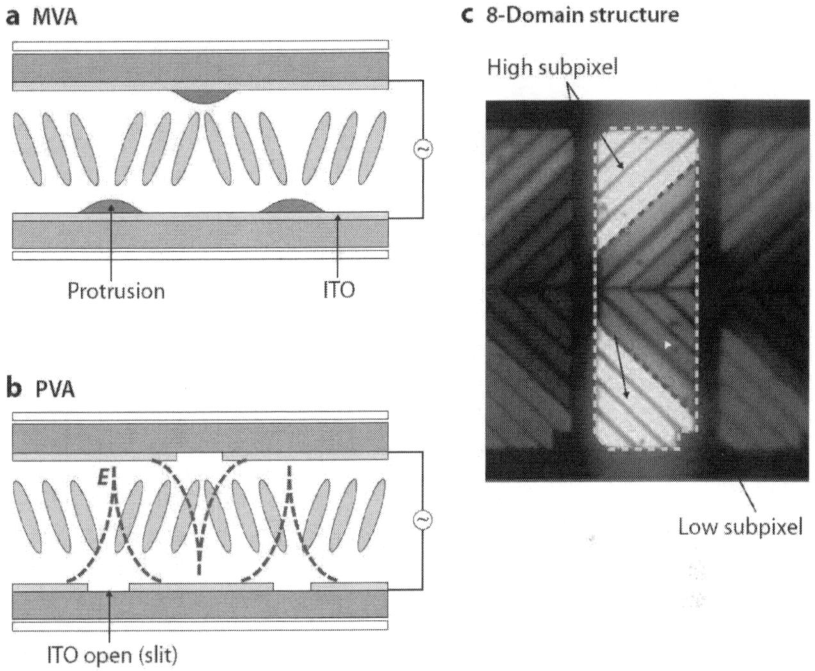

Figure 3. Schematic diagrams of the (a) MVA and (b) PVA structures. (c) Microscope image of eight-domain structure.

The other complementary technology that has been developed to overcome the poor off-axis image quality for VA panels is the compensation film.

1.4 Compensation films for VA and IPS modes

The IPS mode has excellent intrinsic image quality, but diagonal off-axis black quality is an area requiring improvement. The maximum polar angle at which the contrast ratio between white and black states is greater than 10:1 is about 60° off-axis in the diagonal direction. The poor diagonal off-axis black quality of the IPS mode can be improved through the use of a biaxial compensation film known as a 'z-plate', which has a refractive index ellipsoid characterized by $n_x > n_z > n_y$ and $(n_x - n_z)/(n_x - n_y) \approx 0.5$ (z axis is normal to the film, x and y axes are on the film plane).[33] However, the z-plate film is difficult to manufacture and hence expensive, and so most IPS panels are currently produced without the film. Instead, a low-retardation triacetyl cellulose (TAC) polarizer is used. The TAC film has almost no birefringence, and improves the color shift that occurs under diagonal off-axis viewing in the IPS mode.[34]

Although the off-axis diagonal black quality of VA panels is even worse than that for the IPS mode, the diagonal off-axis angle meeting the 10:1 contrast requirement can be expanded to more than 89° with the use of an appropriate compensation film. These films are indispensable for VA-mode LCDs, and have been studied and adopted from the beginning of development of the MVA and PVA modes.[35,36] Various materials and simple structures have been developed for the compensation film, and the films have become much more affordable than the z-plate for the IPS mode. As the structure of the compensation film for the VA mode is simple compared to that used for TN panels, the compensations films for VA panels are even less expensive than the wide-viewing-angle compensation films now used for TN displays [37].

Compensation films should compensate for optical retardation in the oblique direction associated with the geometry of the LC layers, and also for the geometric deformation of the relative angle between the two polarizers (the angle between the two polarizers is 90° on-axis, but less or more than 90° off-axis).[36,38] For VA-mode LCD panels, a negative C-plate film with a refractive index ellipsoid of $n_x = n_y > n_z$ is used to compensate for the retardation of the LC layer, which has an ellipsoid of $n_x = n_y < n_z$. However, to compensate for the change in relative angle between the polarizers, the refractive index n_x of the compensation film should be different fromn_y. This has been achieved by combining films, such as a combination of an A-plate ($n_x > n_y \approx n_z$) and a C-plate, two biaxial plates ($n_x > n_y > n_z$), one biaxial plate, or a biaxial plate and a C-plate.

2. MOVING IMAGE QUALITY

2.1 Factors affecting moving image quality

The quality of moving images is another weak point of LCDs, which generally have slower response times than other display devices.[16,17] This relatively poor moving image performance is due to the slow molecular motion of liquid crystal, which causes blurring of the moving picture, and the 'holding type' method of display, by which each frame is held on the screen until refreshed with the subsequent frame. To improve the moving image quality, these two different factors should be taken into account independently.

2.2 Response time

The molecular motion of liquid crystal is slow (tens of milliseconds) compared to the frame refresh rate, which results in blurring in the moving image. To achieve faster molecular motion, various factors should be optimized, including the LC material, cell thickness, pixel design, electrode shape and driving scheme.

Over the last 15 years, new LC materials having low viscosity, which is a key factor in determining the response time, have been developed.[39] The lowest viscosity of current IPS liquid crystal (with positive dielectric anisotropy) is about 50–60 mPa s, whereas that for VA liquid crystal (with negative dielectric anisotropy) is about 100–110 mPa s. However, as the distance between opposing electrodes in VA cells is the same as the cell thickness (< 4 μm) and shorter than that between the interlaced electrodes in an IPS cell (ca. 10 μm), the effective response times of these two types of cells are very similar. The IPS mode has a less variable gray-to-gray response time than the VA mode, which is an advantage for the IPS mode.[40] The elastic constant is also an important factor contributing to the response time.[41]Thinner cells impart stronger restoring force due to higher stress, resulting in faster off-response times.[42,43] Technologies for a thinner cell gap have therefore been developed to improve response times; some LCDs now have cells of as thin as 3 μm.

To further improve the response time of LCDs, overdriving is now commonly used in many high-end displays.[44,45,46] The response time (τ) of liquid crystal is inversely proportional to the square of the applied electric field, as given by $\tau \propto 1/(V_{app} - V_{th})^2$ where V_{app} is the applied cell voltage and V_{th} is the threshold voltage of the LC. The response time is therefore sensitive to the strength of the effective electric field applied across the LC layer. However, the TFT-LCD has a weak point in this regard. The electric charge (Q) in the pixel is preserved as both storage capacitance (C_{st}) and LC capacitance (C_{lc}) until the charge is refreshed by the next gate signal. The apparent dielectric constant (ε) of the LC layer along the electric field changes during molecular movement, and the actual voltage (V) on the pixel drops gradually, since $V = Q/(C_{st} + C_{lc})$ and $C_{lc} \propto \varepsilon$. Due to this voltage drop, the LC molecules cannot rotate from the initial to the final states within a single frame, and successive recharges of voltage are required in order to realize the final state over the next several frames. Hence, the transmittance of an LC cell undergoes a step-wise increase over time, termed a 'cusp', resulting in a delay in the response time, as shown in Figure 4.[44,45] To remove the cusp by compensating for the dynamic capacitance change, a higher voltage with respect to the target voltage is applied for turn-on, and a lower voltage is applied for turn-off (Figure 4). This is the concept behind overdriving. In reality, the overdrive voltage is determined so as to achieve the fastest response time without introducing side-effects such as ghosting or tailing of the image. [46]

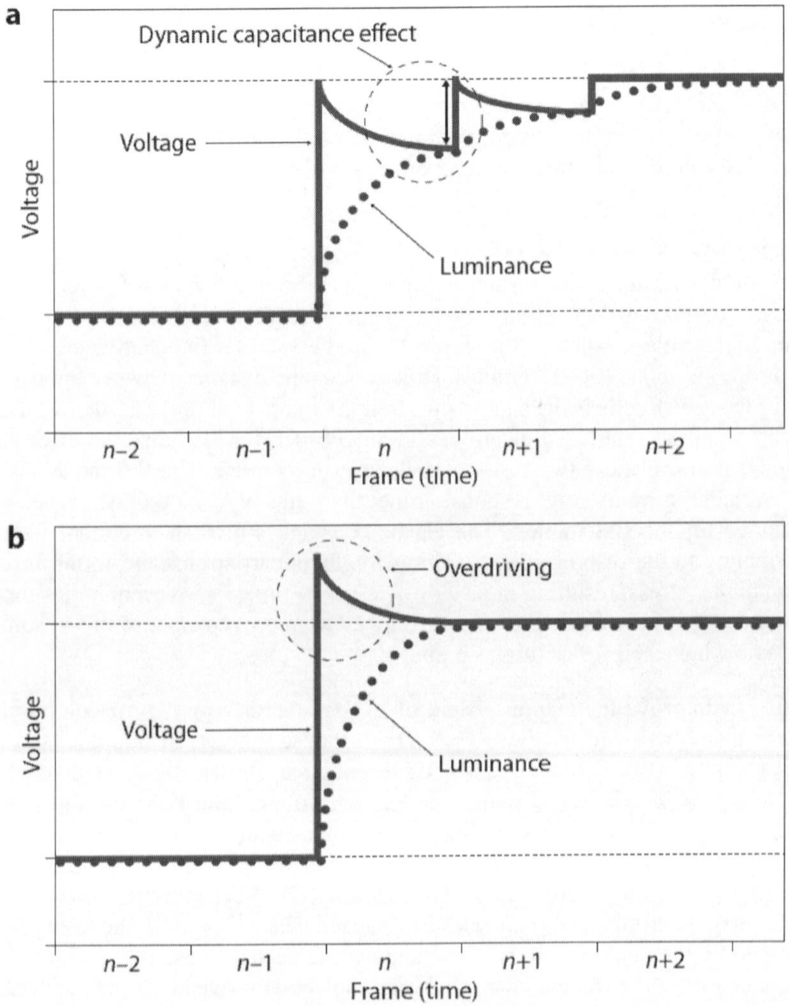

Figure 4. Schematic diagram illustrating the concept of 'overdriving'. (a) The 'cusp' effect is a step-wise increase in LC cell transmittance that results in a delay in response time. (b) To compensate for the dynamic capacitance change, a higher voltage is applied for turn-on, and a lower voltage is applied for turn-off.

In VA cells, control of the tilt direction of molecules to prevent the development of an unstable Schlieren texture is also important for improving the response time. One method for suppressing the formation of texture is the application of a pre-tilt voltage prior to application of the conventional overshoot voltage.[47] The pre-tilt signal helps the LC molecules tilt over to the desired direction without forming a Schlieren texture, and improves the response time significantly. Another method for suppressing the formation of an unstable texture is the use of an electrode with a fine micro-slit pattern. The molecules tend to align parallel with the micro-slits, suppressing azimuthal movement of the LC molecules.[48,49] Photo-alignment and reactive mesogens are also being developed to induce a pre-tilt angle.[50,51]

2.3 Holding-type display problem

CRT displays are composed of phosphor-based pixels that 'flash' for a sub-millisecond duration once each frame under excitation by a scanning electron beam. In contrast, a pixel in an LCD maintains a constant luminance for the entire duration of the frame, known as 'holding'. The LCD image therefore remains unchanged for the entire frame until replaced by a new image signal, resulting in a discretely changing image, that is, a static image that is held for one frame and then changes abruptly in the next frame. However, the human eye moves continuously and linearly. The difference between the trajectory of the eye and the movement path of the image on the display causes the viewer to perceive the moving image as being blurred (Figure 5(a)), no matter how fast the response time of the liquid crystal.[52,53,54,55,56]

To solve the problem of motion blurring, two independent methods have been developed and adopted successfully in high-end LCDs. One method, impulsive driving,[57,58,59] imitates the display mechanism of the CRT by inserting a black image between successive display frames, either directly or by blinking the backlight lamp in part of the panel (Figure 6(a,b)). The longer the duration of the dark state, the better the quality of the moving image. However, both techniques reduce the luminance of the LCD and introduce flicker.

The other method, known as the high refresh rate method,[60,61] reduces blurring without introducing flicker or reducing luminance or contrast by multiplying the frame frequency and inserting one or more interpolated images between successive frames (Figure 6(c)). This method is widely used in commercial displays. The effects of the impulsive driving and high refresh rate methods are illustrated in Figures 5(b)and (c). The difference between the trajectory of the eye and that of the image is reduced significantly. Recently, a 240 Hz driving method, in which three interpolated images are inserted between successive frames, has achieved similar moving image performance to that of CRT displays. Table 1 lists the measured motion picture response times (MPRTs) for various display rates with and without impulsive driving of the backlight.[55,56] This measurement is the quantitative response time for moving images considering both the LC molecular motion and the holding image effect. For comparison, a CRT display has an MPRT of about 5 ms.

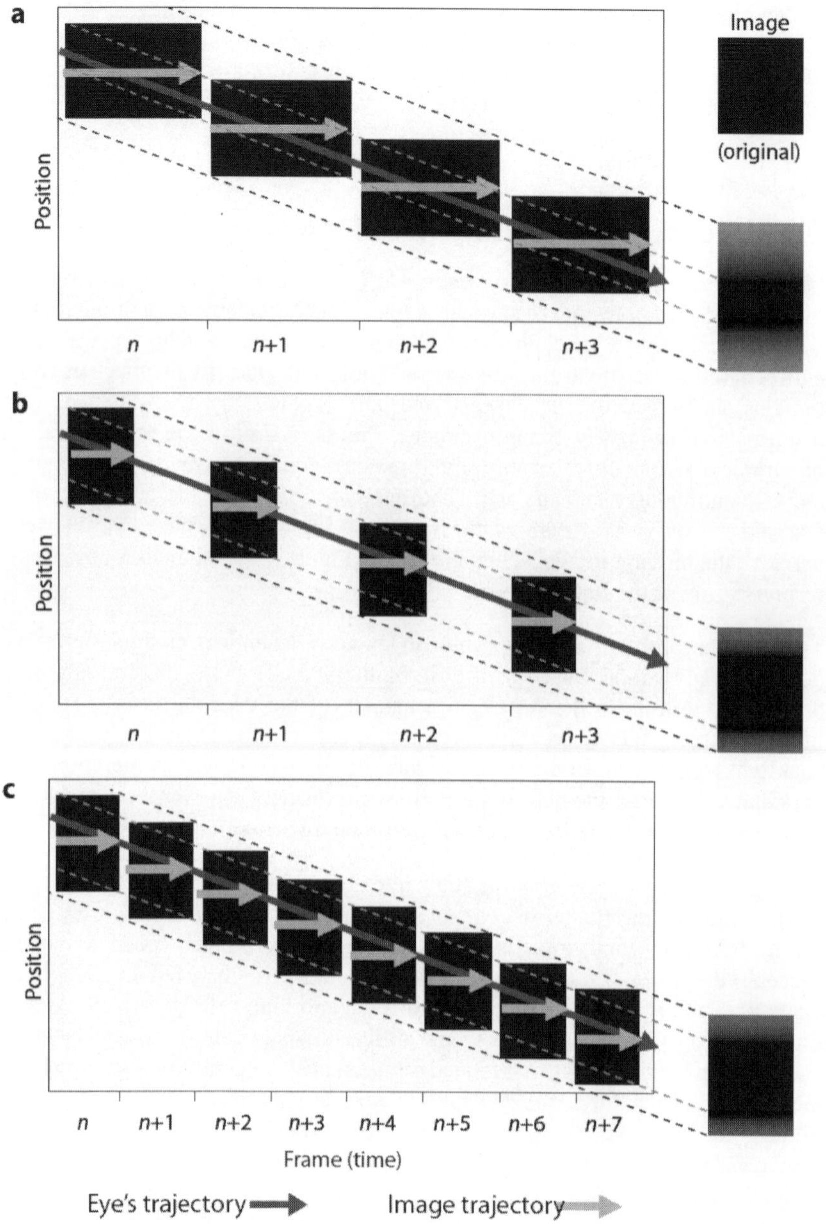

Figure 5. Mechanisms of image blurring in holding-type displays. (a) Conventional motion. (b) Impulsive driving. (c) High refresh rate driving.

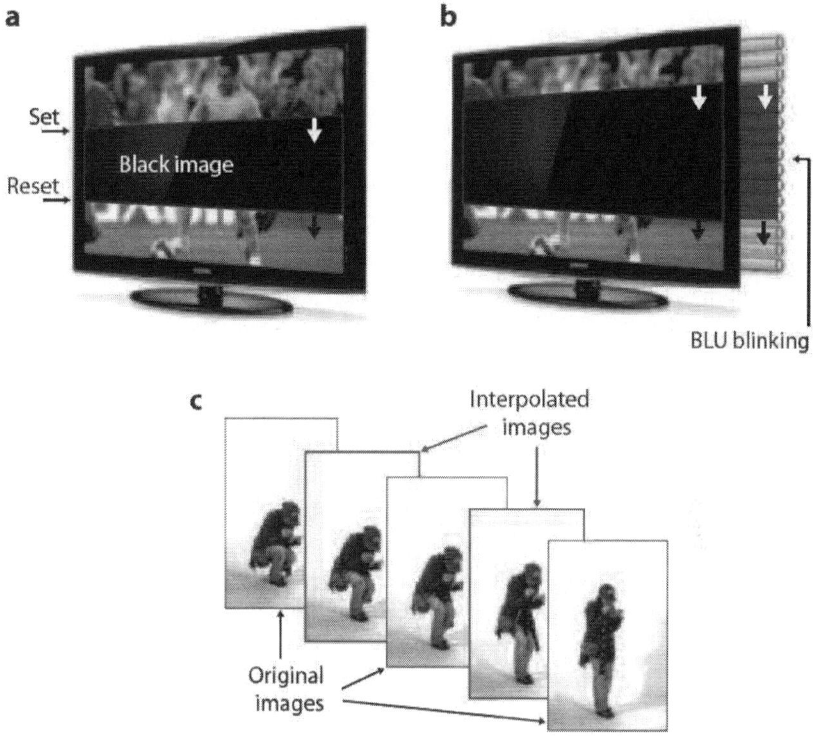

Figure 6. Technologies for enhancing motion picture response time in holding-type displays. (a) Black image insertion. (b) Blinking (scanning) backlight unit. (c) High refresh rate method.

Table 1 . Measured MPRTs for an S-PVA LCD television panel.

Frame frequency	No impulsive driving	With impulsive driving
60 Hz	18.5 ms	9.6 ms
120 Hz	8.7 ms	6.5 ms
240 Hz	5.0 ms	3.9 ms

3. CONTRAST RATIO

The contrast ratio for LCDs, defined as the ratio of white luminance to black luminance, is one of the most important factors contributing to clear and vivid image quality.[29] For example, when watching a movie on a home display, black luminance greater than 0.1–0.2 Cd/m^2 can deteriorate the perceived quality of dark images. An LCD television typically has a brightest luminance of about

500 Cd/m^2 and a contrast ratio of 1,000:1, meaning that the black luminance is 0.5 Cd/m^2, which is not sufficiently dark to represent a vivid black image.

For IPS panels in particular, the planar orientation of LC molecules makes it rather difficult to prevent light leakage. In the dark state, LC molecules are uniformly aligned parallel to the direction of the polarizers, which is achieved by rubbing of the alignment layer. However, such a mechanical rubbing process cannot ensure the perfectly uniform alignment of molecules, and random defects deteriorate the black quality. It has recently been reported that a contrast ratio of 1,500:1 has been achieved for IPS-mode panels through the use of an improved rubbing process and materials with low light scattering properties.[62,63]

In VA-mode panels, a very dark state can be achieved because there is no residual retardation in the field-off state.[61] The micrometer-high protrusions in MVA-mode cells could cause light leakage due to the slight tilting of the LC molecules. Hence, the protrusions should have a gentle inclination to minimize light leakage. The PVA mode, which has an ITO slit, has very little light leakage owing to its flat surface geometry, and recently contrast ratios of around 10,000:1 have been reported at several international conferences.

To obtain very high contrast ratios, both the LC mode and the scattering property of the materials are very important. Light leakage by scattering in the various layers of an LCD panel can be as large as that caused by residual retardation in the LC layer.[64,65,66] Light scattering from the color filter layer in particular has the largest contribution to light scattering in an LCD.[66,67,68] To minimize scattering from the color filter, technologies for reducing the size of color pigments and increasing the degree of pigment dispersion have been developed. Recently, small color pigments of around 60 nm in diameter have been developed. However, as the size of the pigment molecules decreases, the aggregation of molecules becomes stronger, which limits the minimum pigment size. To overcome this aggregation, color filter materials containing dye colorants are being re-investigated. Another important factor for obtaining high contrast ratios is the material of the polarizer. The ability of the polarizer to polarize light is known to be important with regard to the contrast ratio,[69] and a number of technologies have been studied and developed to improve the polarizer, including use of the poly(vinyl alcohol)-iodine complex, and the introduction of low-haze surface treatment and low-light-scatter compensation films.

Conventionally, the contrast ratio is measured under dark conditions. Recently, however, the 'ambient' contrast ratio (measured under ambient lighting conditions) has become more important, particularly for outdoor applications such as mobile displays and advertisement boards.[70,71,72] Compared to PDPs or CRT displays, LCDs have better contrast ratios under ambient conditions because, unlike PDPs and CRTs, the LCD does not use phosphor, which emits light when excited by ambient incident light.[73] To further improve the ambient contrast ratio for LCDs, suitable surface treatments for reducing reflections on the surface of the polarizer are currently under vigorous study.[72]

Another important technology for improving the contrast ratio is the dynamic contrast ratio method, which involves localized dimming of the display using an image-synchronized backlight system. The basic concept of local dimming is shown inFigure 7. By reducing and increasing the brightness of the backlight unit in dark and bright areas of the image, respectively, the quality of the image and the apparent contrast ratio can be much improved. By dimming the backlight in dark areas, dynamic contrast ratios of 150,000:1 and 1,000,000:1 can be achieved for cold cathode fluorescent lamp (CCFL) and light-emitting diode (LED) backlight technologies, respectively. However, local dimming can introduce artifacts such as the degeneracy of gray levels in the dark or bright state. In particular, the difference in black level between dimmed and non-dimmed areas can be perceptible. Such artifacts can be minimized when the panel has a high contrast ratio in the conventional sense. An appropriate algorithm for optimizing image data and backlight luminance should be adopted. Table 2 shows the improvement in the contrast ratio with the refinement of these technologies. The increase in the ambient contrast ratio over the last few years is primarily due to improvements in polarizer surface treatment.

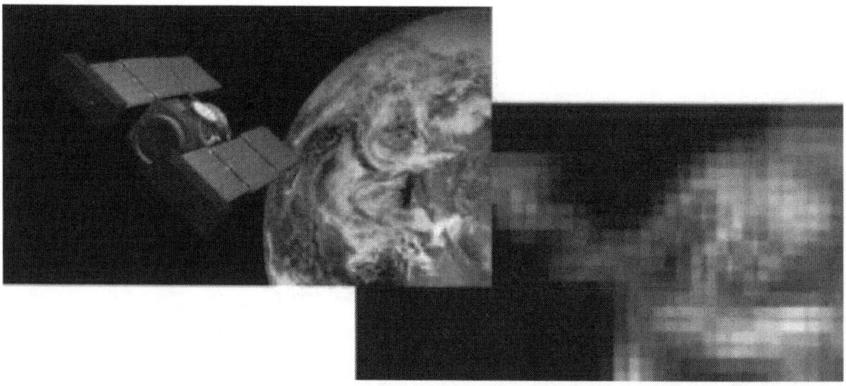

Figure 7. Local dimming method for backlights consisting of a matrix of LEDs.

4. COLOR PERFORMANCE

Accurate color reproduction is another performance factor for high-end LCDs. The ability of an LCD panel to display a continuous and vivid color spectrum is determined by the absorption spectra of the color filters forming the panel and the spectral character of the backlight. The color filter material is composed of a basic resin and colorants, and the quality of colorants is the most important factor in achieving superior color performance.[67,68] Dye colorants were used originally in LCDs, but were later replaced with pigment colorants, which offer greater heat resistance and durability. However, pigments form molecular aggregates of several tens to a few hundred nanometers in size, which causes increased light scatter and a consequent reduction in the contrast ratio. Hence, dye colorants have again attracted the interest of researchers due to their superior qualities in terms of contrast ratio and transmittance. Recent technical developments have demonstrated good progress toward overcoming the lower heat resistance and durability of dye materials.

Table 2 Improvement in measured contrast ratios for S-PVA LCD television panels.

	Year			
Contrast ratio	**2003**	**2006**	**2008**	**2009**
Conventional	1,000:1	2,000:1	5,000:1	8,000:1
Ambient*	600	1,250	1,600	1,800
Dynamic	5,000:1	10,000:1	50,000:1	150,000:1
Polarizability	99.988	99.993	99.996	99.998

*Under 200 lx external luminance

The maximum color reproducibility currently achievable for an LCD panel based on control solely using a color filter layer is about 90–95% of the National Television System Committee (NTSC) color scheme. This is higher than that for both PDPs and CRT displays, but less than the 120% of NTSC achievable by OLED-based displays. By combining a thick color filter with a LED backlight system, the color reproducibility can be extended to 150% of the NTSC standard, which is at present the best color reproducibility available for flat panel displays.

The standard red-green-blue (sRGB) color space is known as 'true color' in display systems,[74] and most recent LCD monitors and LCD televisions cover the sRGB color space, which corresponds to 71% of the NTSC standard. As the technologies for digital cameras and printers improve, sRGB has become insufficient for reproducing the vivid images taken by high-end digital cameras. New standards such as the Adobe and xvYCC color spaces, which cover approximately 100% of the NTSC color space, have been proposed, and high-end LCD monitors and televisions covering these color spaces are currently in development.[74,75]

5. ENVIRONMENTAL TECHNOLOGIES AND HIGH-APERTURE-RATIO PIXEL STRUCTURE

Environmental issues have recently become more important in the LCD industry, and 'green' technologies are now of particular interest. Such technologies typically involve reducing the reliance on hazardous materials and lowering power consumption.

The Electronic Products Environmental Assessment Tool (EPEAT) offered by the Green Electronics Council in the US designates a list of hazardous materials that should be omitted from electronic products (see http://www.epeat.net/), such as cadmium, mercury, arsenic, lead, hexavalent chromium and certain brominated flame retardants. Hence, the development of new substitutes for these hazardous materials without degrading image quality is an important aspect of LCD research and affects most LCD technologies, including the liquid crystal, color filter and florescent lamp.

Low energy consumption is also an important goal for LCDs, for both consumers and industry, in order to reduce the environmental burden of these devices and to extend the operating time of mobile products. The Energy Star 5.0 standard, developed by the Environmental Protection Agency and the Department of Energy in the US, will come into effect in October 2009. This new standard enacts tighter regulation of maximum power consumption for LCDs, defined according to size and resolution (see http://www.energystar.gov/index.cfm?c=about.ab_index). More than 60% of the total power consumption for a notebook LCD panel and more than 80% of that for an LCD monitor or television is attributable to the backlight unit (BLU). High-efficiency backlight systems and high-transmittance LCD panels are therefore essential technologies for achieving low power consumption. LED BLUs consume about 30% less energy than CCFL BLUs, and are thus a good light source for low power consumption applications. Light-recycling film, which re-uses light by transmitting light polarized parallel to the film and reflecting light with perpendicular polarization,[76] is another power-saving technology that can enhance the luminance of the LCD panel by about 30%.

Backlight-dimming technology, originally developed to enhance the dynamic contrast ratio, also reduces power consumption.[61,77,78] The backlight luminance in this case is varied according to the maximum gray level of the displayed image. For example, if the maximum gray level of the image is half the full white level, the backlight luminance is dimmed to 50% of full luminance by reducing the electric current supplied to the lamps, and the image data are modified to effectively double the panel transmittance so as to achieve the same luminance as the original image. In the case of edge-type fluorescent BLUs, which are installed on one or two edges of the panel, dimming is performed globally, changing the luminance uniformly over the entire display area. This method can reduce power consumption by about 8%. To further increase energy efficiency, the panel can be divided into segments, and the luminance of each segment can be controlled independently.[78] In the case of direct CCFL BLUs, which consist of many long florescence lamps located across the entire panel, and direct LED BLUs, which consist of a matrix of LED lamps on the backlight plate, one-dimensional and two-dimensional local dimming can be achieved, respectively, by controlling the luminance of each CCFL or LED lamp independently. The average power saving for these dimming technologies can reach about 30% and 50%, respectively. The use of LEDs has thus become more important not only for improving image quality but also for reducing power consumption.

However, LED and light-recycling films remain comparatively expensive and are thus only available in expensive high-end products. LCD manufacturers have therefore been attempting to increase the transmittance of the LCD panel itself, which provides power savings without increasing cost.

IPS panels consist of an interlaced array of metallic pixel and common electrodes. As the electrodes block light, the presence of the electrodes reduces the aperture ratio and transmittance. Advancements in manufacturing and processing equipment have reduced the width of the electrode significantly, and the aperture ratio and transmittance of IPS panels has been improved considerably from early models. High-aperture-ratio structures incorporating an organic insulation layer have also been developed.[79] The organic layer is placed on top of the data bus line layer, and a common electrode on the organic layer covers the data bus line, preventing light leakage near the bus line. The width of the black matrix can be narrowed significantly by this approach. The FFS mode uses transparent electrodes for both pixel and common electrodes and therefore has much higher transmittance than the conventional IPS mode. For this reason, the application of the FFS mode in mobile devices and televisions is increasing.[24]

Up until the late 1990s, the aperture ratio and transmittance of VA panels were much higher than those for IPS panels because the VA mode uses transparent electrodes. However, the adoption of the eight-domain pixel structure for the VA mode to improve off-axis image quality has resulted in lower transmittance due to a reduction in the aperture ratio caused by the insertion of additional pixel structures. This has diminished the transmittance advantage of the VA mode over the IPS mode. To improve the transmittance of the VA mode, high-aperture-ratio structures using a 'color filter on array' process, and the polymer-stabilized VA mode have been developed.[48,49] In the former technology, the color filter and black matrix layers are placed between the data and pixel electrodes on the TFT substrate. This structure narrows the linewidth of the black matrix, and also ensures that there is no misalignment between the top and bottom glass layers. The latter technology[48,49]eliminates the need for a patterned ITO electrode in constructing the multi-domain pixel structure, thereby increasing the aperture ratio. This is achieved by using liquid crystal containing a small amount of a reactive mesogen. The LC molecules are aligned according to micro-slits on the pixel electrode, and complete alignment is achieved after application of an electric field for a few seconds. Exposure to ultraviolet light then produces a non-zero pre-tilt angle of the LC molecules in the absence of an electric field. The ITO patterns or protrusions are therefore no longer necessary for the multi-domain structure, and a high aperture ratio can be achieved. A small number of reactive mesogen materials have been developed to date. Photo-alignment technology, developed to improve transmittance, is a major recent achievement that has been applied in a small number of commercial products.[50,51]

A four-subpixel RGBW structure is another candidate technology for reducing power consumption.[80,81,82] In this technology, a white pixel is added to the conventional three-subpixel RGB structure to enhance the luminance of the panel by as much as 150%. An image-processing algorithm is required in this case to minimize the degradation of color reproducibility caused by the addition of a white subpixel.

Field-sequential color LCD technology is a new development that theoretically provides a three-fold increase in transmittance by removing the color filter layers from the LCD panel.[83] In this technology, a pixel three times larger than that in usual LCD panels displays red, green and blue signals sequentially in synchronization with red, green and blue backlights. To realize this technology, the response time of the LCD panel should be shorter than 1–2 ms, and to reduce color separation at the edge of a moving image, the frame frequency should be higher than 180 Hz.

6. CONCLUSION AND PROSPECTS FOR LCD TECHNOLOGY

The LCD is the culmination of a set of elemental technologies ranging from materials—including condensed, organic, inorganic and polymer materials—to semiconductors, electronics, thin films and LEDs. The advancement of LCD technology therefore influences, and is influenced by, many other sciences and industries, particularly the materials and equipment industries. Over the past few decades, a wide variety of new LCD technologies have been introduced, including numerous LC modes, driving schemes and panel structures, yet only a few have been successfully adopted in commercial products. This does not mean that only a few of the technologies are useful, but rather that there are many technologies that are still waiting for suitable applications. The success of certain technologies is critically dependent on advances in materials science. For example, the development of liquid crystal with negative dielectric anisotropy allowed the development of the VA modes, and the discovery of reliable alignment polymers solved the serious image sticking problem for the IPS mode. Large-area glasses and processing equipment such as 'one-drop filling' equipment have also reduced the manufacturing cost of LCDs dramatically, while the development of low-light-scattering materials, high-quality color pigments and functional films has improved the image quality of LCDs, which in many cases are now superior in quality to other types of displays.

New materials have thus afforded new applications and opened new markets for LCDs, and in the same way, future LCD technologies can also be expected to be dependent on advances in materials. For example, low-cost three-dimensional displays based on time-division stereoscopic technology, which requires high frame frequency and hence fast LC response time, is dependent on the development of low-viscosity LC materials. However, the development of cost-effective and reliable lenticular films or patterned retardation films could also lead to technologies that might dominate the three-dimensional display market. Similarly, the flexible display market might be expanded with the development of new organic films and organic TFT technologies. There are many such LCD technologies that have been developed but which are awaiting complimentary advances in materials research; photo-alignment technology, blue phase mode, cholesteric LCD and printing technology as a replacement to photolithography are just a few examples.

REFERENCES

1. F. Reinitzer, *Monatsh. Chem.* **9**, 421 (1888). |
2. F. Reinitzer, *Ann. Physik.* **27**, 213 (1908).
3. B. Bahadur, ed., *Liquid Crystal – Applications and Uses* **Vol. 1**, World Scientific, Singapore (1990).
4. M. Schadt, W. Helfrich, *Appl. Phys. Lett.* **18**, 127 (1971). |
5. T. J. Scheffer, J. Nehring, *Appl. Phys. Lett.* **45**, 1021 (1984). |
6. E. P. Raynes, *Electron. Lett.* **10**, 10 (1974).
7. E. P. Raynes, *Electron. Lett.* **10**, 141 (1974). |
8. Miyaji, M. Yamaguchi, A. Toda, H. Mada, S. Kobayashi, *IEEE Trans. Electron Devices* **ED-24**, 811 (1977).
9. H. Fukuro, S. Kobayashi, *Mol. Cryst. Liq. Cryst.* **163**, 157 (1988).
10. M. F. Schiekel, F. Fahrenschon, *Appl. Phys. Lett.* **19**, 391 (1971).
11. J. F. Clerc, M. Aizawa, S. Yamauchi, T. Yamamoto, Y. Kondo, *Proc. Japan Display 1989*, 762 (1991).
12. F. J. Kahn, *Appl. Phys. Lett.* **20**, 199 (1972).
13. R. A. Soref, *Appl. Phys. Lett*. **22**, 165 (1973).
14. T. P. Brody, J. A. Asars, G. D. Dixon, *IEEE Trans. Electron Devices* **ED-20**, 993 (1972).
15. T. P. Brody, *J. Soc. Inf. Display* **4**(3), 113 (1996).
16. S. Mikoshiba, *Soc. Inf. Display Seminar Lecture Notes* **1**, M-4 (1997).
17. D. Parker *et al.*, *Soc. Inf. Display Int. Symposium Digest of Technical Papers***29**, 15 (1998).
18. P. J. Bos, K. R. Koehler, *Proc. Japan Display 1983* **113**, 478 (1983).

19. Y. Yamaguchi, T. Miyashita, T. Uchida, *Soc. Inf. Display Int. Symposium Digest of Technical Papers* **24**, 277 (1993).

20. N. A. Clark, S. T. Lagerwall, *Appl. Phys. Lett.* **36**, 899 (1980).

21. R. Kiefer, B. Webber, F. Windscheid, G. Baur, *Proc. Japan Display 1992*, 547 (1992).

22. M. Oh-e, K. Kondo, *Appl. Phys. Lett.* **67**, 3895 (1995).

23. K. Kondo, K. Kinugase, N. Konishi, H. Kawagami, *Soc. Inf. Display Int. Symposium Digest of Technical Papers* **27**, 81 (1996).

24. S. H. Lee, S. L. Lee, H. Y. Kim, *Appl. Phys. Lett.* **73**, 2881 (1998).

25. M. Oh-e, *Liquid Crystals Today* **10** (2001).

26. H. Wakemoto *et al.*, *Soc. Inf. Display Int. Symposium Digest of Technical Papers* **28**, 929 (1997).

27. Takeda *et al.*, *Soc. Inf. Display Int. Symposium Digest of Technical Papers***29**, 1077 (1998).

28. K. H. Kim *et al.*, *Proc. Asia Display 1998*, 383 (1998).

29. S. S. Kim, B. H. Berkeley, K. H. Kim, J. K. Song, *J. Soc. Inf. Display* **12**(4), 353 (2004).

30. S. S. Kim, B. H. Berkeley, J. H. Park, T. Kim, *J. Soc. Inf. Display* **14**(2), 127 (2006).

31. S. B. Park *et al.*, *Soc. Inf. Display Int. Symposium Digest of Technical Papers***38**, 1252 (2007).

32. Y.-P. Huang *et al.*, *Soc. Inf. Display Int. Symposium Digest of Technical Papers***38**, 1010 (2007).

33. Y. Saitoh, S. Kimura, K. Kusafuka, H. Shimizu, *Soc. Inf. Display Int. Symposium Digest of Technical Papers* **29**, 706 (1998).

34. T. Yasuda *et al.*, *SSoc. Inf. Display Int. Symposium Digest of Technical Papers***37**, 1169 (2006).

35. K. Ohmuro, S. Kataoka, T. Sasaki, Y. Koike, *Soc. Inf. Display Int. Symposium Digest of Technical Papers* **28**, 845 (1997).

36. J. Chen *et al.*, *Soc. Inf. Display Int. Symposium Digest of Technical Papers* **29**, 315 (1998).

37. H. Mori, Y. Itoh, Y. Nishiura, T. Nakamura, Y. Shinagawa, *Proc. AM-LCD 1996*, 189 (1996).

38. J. E. Bigelow, R.A. Kashnow, *Appl. Opt.* **19**, 2090 (1977).

39. C. S. Lim, J. H. Lee, C. H. Oh, H. C. Choi, *Proc. Int. Meeting on Information Display 2003*, 68 (2003).

40. M. Ohta *et al.*, *Soc. Inf. Display Int. Symposium Digest of Technical Papers* **28**, 86 (1997).

41. C. S. Lim *et al.*, *Proc. Int. Meeting on Information Display 2004*, 26–4 (2004).

42. H. Hasebe, S. Kobayashi, *Soc. Inf. Display Int. Symposium Digest of Technical Papers* **16**, 81 (1985).

43. T. Tanaka, H. Hasebe, S. Kobayashi, *Proc. Japan Display 1986*, 396 (1986).

44. B. W. Lee *et al.*, *Proc. Int. Display Workshops 2000*, 1153 (2000).

45. B. G. Rho, B. K. Yang, S. S. Lee, J. S. Kim, *Proc. Int. Display Workshops 2000*, 1155 (2000).

46. B. W. Lee, D. Sagong, G. Jeong, *Soc. Inf. Display Int. Symposium Digest of Technical Papers* **32**, 1106 (2001).

47. J. K. Song *et al.*, *Soc. Inf. Display Int. Symposium Digest of Technical Papers* **35**, 1344 (2004).

48. S. G. Kim *et al.*, *Appl. Phys. Lett.* **90**, 261910 (2007).

49. M.-H. Wu, R.-C. Yang, *Proc. Int. Display Workshops 2007*, 1669 (2007).

50. M. Schadt, K. Schmitt, V. Kozinkov, V. Chigrinov, *Jpn J. Appl. Phys.* **31**, 2155 (1992). |

51. W.-C. Lee, C.-S. Hsu, S.-T. Wu, *Jpn J. Appl. Phys.* **40**, 5942 (2001).

52. D. Parker, in *Display Systems* pp. 353–364, John Wiley & Sons, UK (1997).

53. T. Kurita, A. Saito, I. Yuyama, *Proc. Int. Display Workshops 1998*, 823 (1998).

54. T. Kurita, *Soc. Inf. Display Int. Symposium Digest of Technical Papers* **32**, 986 (2001).

55. K. Oka, Y. Enami, *Soc. Inf. Display Int. Symposium Digest of Technical Papers* **35**, 1266 (2004).

56. Y. Igarashi *et al.*, *Soc. Inf. Display Int. Symposium Digest of Technical Papers* **35**, 1262 (2004).

57. N. Fisekovic, T. Nauta, H. J. Cornelissen, J. Bruinink, *Proc. Int. Display Workshops 2001*, 1637 (2001).

58. M. A. Klompenhouwer, *Soc. Inf. Display Int. Symposium Digest of Technical Papers* **36**, 1578 (2005).

59. S. Sluyterman, E. P. Boonekamp, *Soc. Inf. Display Int. Symposium Digest of Technical Papers* **36**, 996 (2005).

60. Y. Kuroki, T. Nishi, S. Kobayashi, H. Oyaizu, S. Yoshimura, *Soc. Inf. Display Int. Symposium Digest of Technical Papers* **37**, 14 (2006).

61. S. S. Kim *et al.*, *Soc. Inf. Display Int. Symposium Digest of Technical Papers***38**, 1003 (2007).

62. Mori, R. Oke, K. Kamoshida, T. Asakura, K. Ono, *Proc. EuroDisplay 2005*, 102 (2005).

63. Y. Utsumi *et al.*, *Soc. Inf. Display Int. Symposium Digest of Technical Papers***39**, 129 (2008).

64. W. Heller, M. Nakagaki, *J. Chem. Phys.* **61**(9), 3619 (1974).

65. M. Yoneya, Y. Utsumi, Y. Umeda, *J. Appl. Phys.* **98**, 016106 (2005).

66. X. Li, A. Taflove, V. Backman, *Optics Lett.* **30**(8), 902 (2005).

67. Y. Utsumi, I. Hiyama, Y. Tomioka, K. Kondo, S. Matsuyma, *Jpn J. Appl. Phys.***46**, 1047 (2007).

68. L. E. McNeil, R. H. French, *J. Appl. Phys.* **89**(1), 283 (2001).

69. Y. Yoshimi, *Proc. Flat Panel Display Int. 2004*, 146 (2004).

70. C. F. Chao, W. Cheng, *Inf. Display* **37**, 322 (2006).

71. S.-C. Jeng *et al.*, *Soc. Inf. Display Int. Symposium Digest of Technical Papers***38**, 315 (2007).

72. E. F. Kelley, M. Lindfors, J. Penczek, *J. Soc. Inf. Display* **14**(12), 1019 (2006).

73. S. H. Yoo, T. J. Kweon, H. Kwon, E. G. Heo, *Soc. Inf. Display Int. Symposium Digest of Technical Papers* **37**, 1217 (2006).

74. P. Green, L. W. MacDonald, *Colour Engineering: Achieving Device Independent Colour*, John Wiley & Sons, UK (2002).

75. T. Matsumoto *et al.*, *Soc. Inf. Display Int. Symposium Digest of Technical Papers* **37**, 1130 (2006).

76. N. Guselnikov, P. Lazarev, M. Paukshto, P. Yeh, *J. Soc. Inf. Display* **13** (12), 339 (2005).

77. J. H. Stessen, J. G. R. van Mourik, *Soc. Inf. Display Int. Symposium Digest of Technical Papers* **37**, 1249 (2006).

78. T. Shirai, S. Shimizukawa, T. Shiga, S. Mikoshiba, T. Käläntär, *Soc. Inf. Display Int. Symposium Digest of Technical Papers* **37**, 1520 (2006).

79. J.-S. Lin, K.-H. Yang, S.-H. Chen, *Jpn J. Appl. Phys.* **44**, 6178 (2005).

80. B. W. Lee *et al.*, *Soc. Inf. Display Int. Symposium Digest of Technical Papers***34**, 1212 (2003).

81. L. Wang, Y. Tu, L. Chen, K. Teunissen, I. Heynderickx, *Soc. Inf. Display Int. Symposium Digest of Technical Papers* **38**, 1142 (2007).

82. C. H. Brown Elliott *et al.*, *J. Soc. Inf. Display* **11**(1), 89 (2003).

83. L. Arend, J. Lubin, J. Gille, J. Larimer, *Soc. Inf. Display Int. Symposium Digest of Technical Papers* **25**, 201 (1994).

ABOUT THE AUTHORS

Kyeong Hyeon Kim

Kyeong Hyeon Kim received his BS and MS degrees at Korea University and the Korea Advanced Institute of Science and Technology in 1984 and 1986, and earned his PhD degree in organic material engineering at the Tokyo Institute of Technology, Japan, in 1994. He conducted postdoctoral research on ferroelectric liquid crystals at the University of Colorado, USA, before joining Samsung Electronics in Korea. His work on liquid crystal technology has included innovations such as PVA/S-PVA technology, fast-response LC, optical compensation films and new LC modes for wide viewing angle and fast response. He is currently vice president of LC technology in Samsung's LCD research and development team.

Jang-Kun Song

Jang-Kun Song received his BS and MS degrees in physics from Korea University in 1994 and 1996. He joined Samsung Electronics in 1996, where he has been working as project leader on the development of new LCD modes and frontier technologies. He obtained his PhD in electronics engineering from Trinity College Dublin, Ireland, in 2008, where he carried out fundamental research on liquid crystalline materials, publishing 18 papers. He is the main inventor of the PVA and S-PVA modes, and holds about 150 patents including 50 US patents. He is currently principal engineer at Samsung in Korea.

CHAPTER 6

Graphene oxide liquid crystals: synthesis, phase transition, rheological property, and applications in optoelectronics and display

Feng Lin[1], Xin Tong[1], Yanan Wang[1,3], Jiming Bao[3,1] and Zhiming M. Wang[1,2*]

[1]Institute of Fundamental and Frontier Sciences, University of Electronic Science and Technology of China, Chengdu 610054, People's Republic of China.

[2]State Key Laboratory of Electronic Thin Films and Integrated Devices, University of Electronic Science and Technology of China, Chengdu 610054, People's Republic of China.

[3]Department of Electrical and Computer Engineering, University of Houston, Houston, TX 77204, USA.

ABSTRACT

Graphene oxide (GO) liquid crystals (LCs) are macroscopically ordered GO flakes dispersed in water or polar organic solvents. Since the first report in 2011, GO LCs have attracted considerable attention for their basic properties and potential device applications. In this review, we summarize recent developments and present a comprehensive understanding of GO LCs via many aspects ranging from the exfoliation of GO flakes from graphite, to phases and phase transitions under various conditions, the orientational responses of GO under external magnetic and electric fields, and finally Kerr effect and display applications. The emphasis is placed on the unique and basic properties of GO and their ordered assembly. We will also discuss challenges and issues that need to be overcome in order to gain a more fundamental understanding and exploit full device potentials of GO LCs.

KEYWORDS

Graphene oxide liquid crystal, Liquid crystal display, Electro-optical properties, Rheological properties

1. REVIEW

Graphene is an atomically thin carbon material in hexagonal structure and has drawn immense attention due to excellent electrical, thermal, mechanical, and chemical properties and potential device applications [1–5]. Graphene oxide (GO) is synthesized from graphite through wet chemical oxidation and subsequent exfoliation [6–11]. Since GO can be chemically reduced to graphene, initial interest in GO originated from the goal to produce graphene at low cost in large scale [12–15]. It was only after the full development of wet chemical exfoliation of GO that GO LCs were discovered, although LCs of graphene and carbon nanotubes were already observed in chlorosulfonic acid or sulfuric acid [16–18].

Xu and Gao were the first to report nematic phase and isotropic-nematic phase transitions of GO aqueous suspensions in 2011, followed by Kim et al. who investigated the influences of GO flake aspect ratio and NaCl ionic strength on phase transitions [7, 19]. After that, many more detailed studies of basic properties and potential device applications appeared. We believe it is time to review the rapid developments over the past few years and summarize what has been achieved and what the challenges are for future development. We will focus on the unique property of GO and point out how this basic property will affect the characteristic of GO LCs and related device applications. Both strengths and weaknesses of GO LCs will also be discussed.

The organization of this review is as follows: In “Synthesis of Graphene Oxide Liquid Crystals” section, we will talk about various methods that have developed to synthesize GO. In “Phase Properties” section, we will discuss the phase diagram, phase transitions, and their dependence on the mass/volume fraction, size/aspect ratio, salt concentration, and pH value. In “Rheological Properties” section, we will review orientational control and alignment under flow. Birefringence and orientational switch by magnetic and electrical field will be covered in “Magnetic/Electro-Optical Properties” section. We will summarize absorption and fluorescence, shape, and optical anisotropy and optical properties. In “Displays with GO LCs” section, we will discuss the optoelectronic applications. At last, we will conclude the review by conclusion and discussion of challenges and issues for future development.

1.1 Synthesis of Graphene Oxide Liquid Crystals

Graphite oxide was first prepared by Brodie using $KClO_3$ and HNO_3 about 150 years ago [8]. This method was improved by Staudenmaier in 1898 and in 1937 by Hofmann who used concentrated H_2SO_4, HNO_3, and $KClO_3$ to produce highly oxidized graphite. However, this method was time-consuming (about 1 week) and hazardous because of the generation of toxic gases (ClO_2 and NO_x) [9, 10]. In 1958, Hummers reported a new method by replacing HNO_3 and $KClO_3$ with $NaNO_3$ and $KMnO_4$ [11]. Because the oxidation can be completed within 2 h below 45 °C, this Hummers method has been widely used especially after the first mechanical exfoliation of graphene in 2004. It was in the pursuit of producing graphene in large quantity using wet chemical exfoliation that liquid crystals of graphene and subsequently graphene oxide were discovered in 2010 and 2011, respectively [7, 16, 19–23]. Since then, much attention was attracted to the basic properties and new device applications of graphene-based liquid crystals.

1.2 Strategies for Graphite Oxidation

Hummers method has been modified to achieve high qualities of GO with large size or aspect ratio, high yield, and less toxic gases in short time[6]. Marcano and partners used a 9:1 mixture of concentrated H_2SO_4/H_3PO_4 and $KMnO_4$ only to oxidize the graphite flakes with a higher reaction efficiency and less toxic gas [24]. Figure 1(1) shows GO of Hummers method (HGO), improved Hummers method (IGO), and modified Hummers method (HGO+) with additional $KMnO_4$. It can be seen that IGO is more efficient than the other two methods.

To further increase the oxidation efficiency and reduce toxic chemicals, Peng and co-workers developed a green approach to producing GO without heavy metal and toxic gases [25]. As depicted in Fig. 1(2), when K_2FeO_4 is mixed with concentrated H_2SO_4, GO can be synthesized in 1 h at room temperature.

1.3 Exfoliation and Size Control of GO

After oxidation of graphite using the Hummers method, rapid heating and ultrasonic agitation are commonly used to exfoliate graphite oxide into a monolayer [6, 7, 21, 26]. However, these techniques always result in breakage of GO flakes into smaller pieces [6, 20, 27–30]. Aboutalebi and co-workers used large-sized graphite and pre-exfoliation process without sonication, creating ultra-large GO sheets with areas up to 10,000 μm^2 and yield over 80 % [26, 31, 32]. Specifically, with graphite intercalation compounds prepared by stirring the mixture of graphite, H_2SO_4 and HNO_3, then by thermal expansion at 1050 °C and oxidation with $KMnO_4$ at room temperature, GO was obtained in deionized water by gentle hand shaking.

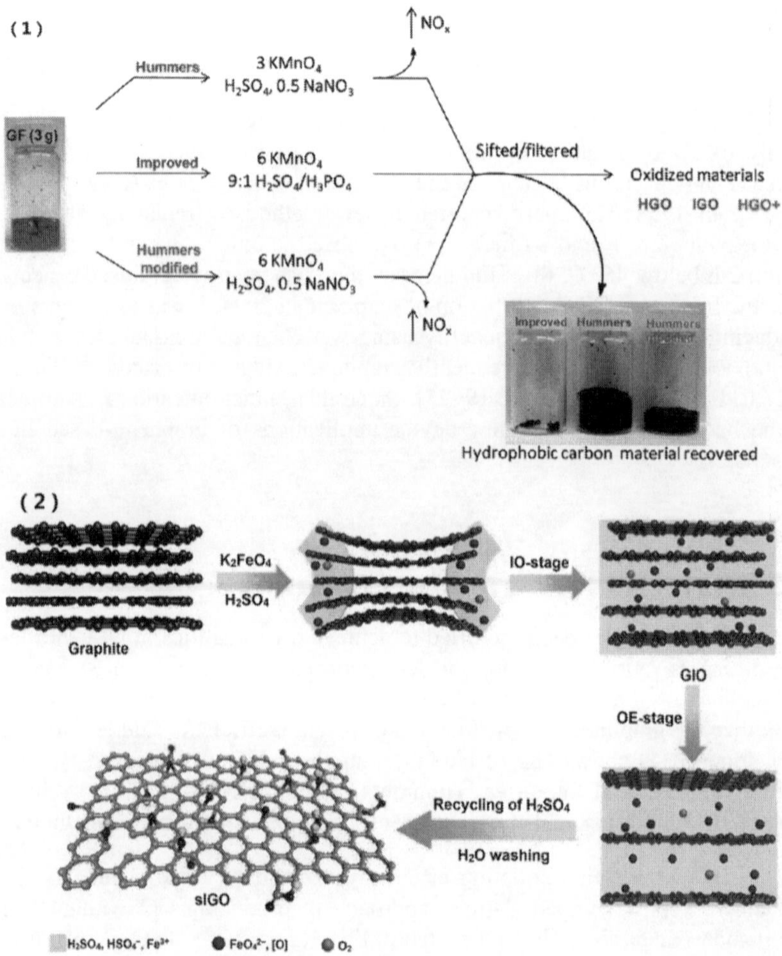

Figure 1 . 1 Oxidation procedures of graphite flakes (GF) by Hummers method, improved Hummers method, and modified Hummers method. The *arrow* and nearby NO_X is gaseous nitric oxide. The *right side* shows generated graphite oxides as well as unoxidized or under oxidized graphite flakes that still exhibit hydrophobic property. Least amount of hydrophobic in the "improved" bottle indicates its highest oxidation efficiency [24]. **2** Synthesis of single layer graphene oxide (slGO) by K_2FeO_4. In intercalation oxidation (IO) stage, the in situ formed oxidants ($FeO_4{}^{2-}$ and atomic oxygen [O]) and O_2 intercalate into graphite layers and form intercalated graphite oxide (GIO). Then, it is further oxidized and exfoliated by O_2 in oxidation-exfoliation (OE) stage. With recycling of H_2SO_4 and water washing, slGO is obtained [25]

With combinations of improvement in exfoliation and organic solvents, a modified strategy of spontaneous exfoliation of graphite oxide in polar aprotic solvents was proposed [30]. A class of organic solvents can be used to exfoliate graphite oxide by simple hand shaking without sonication; they include *N*-methyl-2-pyrrolidone (NMP), dimethyl formamide (DMF), dimethyl sulfoxide (DMSO), dimethyl acetamide (DMAc), and propylene carbonate (PC). Figure 2(1) shows the progress of exfoliation in NMP over time. A complete transformation from graphite oxide to GO is found in 240 s. In contrast, GO cannot be exfoliated in HCl [30]. From exfoliation study, it was also found that GO can be very well dispersed in various organic solvents such as DMF, *N*-cyclohexyl-2-pyrrolidone (CHP), tetrahydrofuran (THF), acetone, and ethanol [33].

When graphite oxide is dispersed in water, a sonication-free exfoliation method was proposed by Ogino et al [34]. This method is a repetitive freeze-thaw cycle which consists of fast freezing graphite oxide solution and then thawing of the frozen solid. As shown in Fig. 2(2), the graphite oxide aqueous solution is first freezing in a liquid N_2 bath and then the sample with ice is thawing in the water bath. After six freeze-thaw cycles, graphite oxide is efficiently exfoliated with minimal fragmentation and yields about 80 % of GO. Further research finds that this method is effective for graphite oxide with high degrees of oxidation (C/O atomic ratios $\leq$2.6) for graphite structure retained would prevent the exfoliation. What is more, faster freezing rate is more efficient for exfoliation and forms high concentration GO dispersions [36]. Another sonication-free exfoliation method injects and maintains CO_2 gas in interlayer of graphite interlayers by freezing and uses the ejection pressure of CO_2 gas to disperse the graphite oxide [35]. In Fig. 2(3), graphite oxide swells in water with increasing the humidity and then high pressure CO_2 is injected and cooled down to −30 °C to induce ice of water to surround the graphite oxide. As the surrounding ice melts, the pressure of ejecting gas spontaneously exfoliated and disperses the GO without sonication. The lateral size of GO exfoliated with above strategies is much larger than with sonication [26, 30, 35, 36]. Besides these sonication-free methods to maintain the lateral size of GO, a multi-step sonication exfoliation was studied without reducing the size of sheets but increasing the yield in contrast to continuous sonication [37].

Moreover, the optimized exfoliation method to achieve large size GO sheets, wide size, and shape distribution could be separated by size selection [38–40]. Some size separation methods based on density gradient ultracentrifuge [40] and pH selective precipitation [39] were put forward, but further purification process is needed for these methods. A facile spontaneous size selection technique without extra additives reveals that with proper concentration of GO dispersed in water, small size sheets and large size sheets will be separated, with small sheets forming isotropic phase and large sheet nematic phase for large size sheets [38].

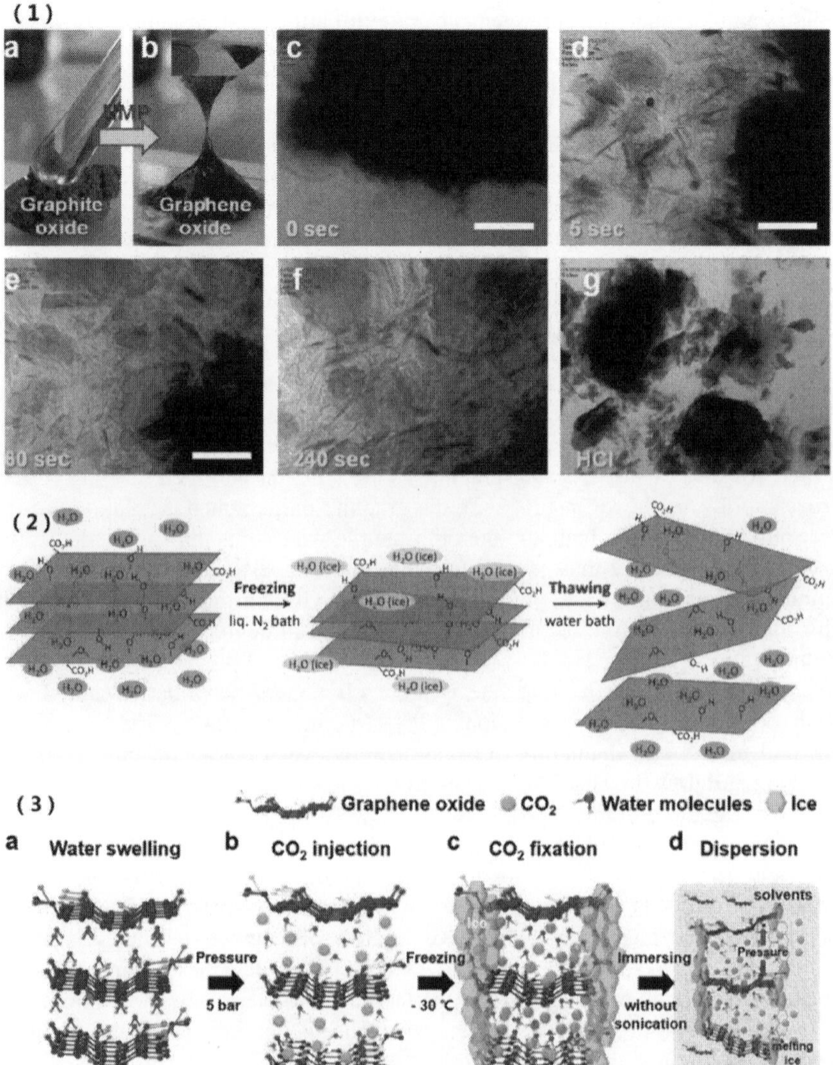

Figure 2. 1 Spontaneous exfoliation of graphite oxide in *N*-methyl-2-pyrrolidone (NMP) solvent. *a* Graphite oxide flakes. *b* GO gel-like dispersion in NMP. *c–f* Graphite oxide exposing to NMP and gradually exfoliates into thin flakes over time. *g* Graphite oxide in HCl solution. The scale bar: 100 μm [30]. **2** A sonication-free exfoliation method with a repetitive freeze-thaw cycle process [34]. **3** Sonication-free exfoliation of graphite oxide by interlayer CO_2 injection. *a, b* graphite oxide swells in water and then high pressure CO_2 is injected. *c* Cooling to −30 °C. *d* Immersed in water, the pressure of ejected CO_2 gas separates GO layers [35]

Specified size of GO can be achieved by controlling starting graphite, oxidation, and exfoliation procedure. The large size of precursor natural graphite favors larger sheets though it is not necessary to form high lateral size GO [41, 42]. In the oxidation process, it is found that oxidation time, volume of oxidants, and oxidation path will also affect the GO particle size [41, 43–45]. Zhang et al. observed that the mean size of GO sheets descends with longer oxidation time and more oxidants [44]. It is also observed that higher oxidation degree with C/O of 2.08 shows larger GO size than it with C/O of 2.63 after same sonication [46]. With the increasing of oxidation degree, the content of oxygen-containing groups like hydroxyl and epoxide groups increased. The increasing oxygen-containing groups further increased the interlayer distance of graphite oxide and finally decreased the van der Waals interlayer interactions. As a result, the graphite oxide was easier to break into small pieces. Furthermore, the oxygen-containing groups decreased the bond energy between carbon atoms. Therefore, C-C bonds and graphite oxide cracked during ultrasonication. The oxidation paths mainly contain cross-planar and edge-to-center ways. The cross-planar oxidation results in periodic cracking of graphene sheets and reduces the lateral size [43]. If ultrasonication is adopted during exfoliation, the size of GO sheets decreases with the increase of sonication time [41, 44, 45]. It can be applied to reduce the GO' particle size when small GO sheets are required in some application.

1.4 Properties of GO LCs

1.4.1 Phase Properties

GO is the oxygenated form of graphene and GO dispersions exhibit stable nematic phase due to the electrostatic repulsive force which originates from hydrolysis of carboxyl and hydroxyl groups on the GO surface. The parameters which have critical influence on the phase transition of GO dispersions include mass/volume fraction [7, 19, 47], size/aspect ratio [7, 47, 48], salt concentration [49, 50], and the pH value [48, 49, 51] of solvents.

The phase transition of GO LCs from isotropic phase to nematic phase was firstly observed with the variation of volume/mass fraction by Xu et al. and Kim et al. [7, 19]. In their research, the influences of aspect ratio and NaCl on phase transition were also investigated. Then, Dan and partners observed the phase transition of giant graphene oxide (GGO) LCs [47]. Tkacz et al. and Zhao et al. reported the systematic research of phase transition based on pH value and salt almost at the same time [48, 51]. Recently, a range of arrested states (glass and gel) of GO dispersion were investigated with various volume fraction and salt concentration by Konkena et al. [50].

The most obvious indication for a liquid crystal is the appearance of birefringence which can be observed between two cross-polarized optical filters. Figure 3a shows that birefringence begins to appear when the mass fraction (f_m) of GO reaches 2.5×10^{-4}, and the birefringence becomes stronger as the concentration of GO increases. It should be noted that these birefringent states are created dynamically by mixing and shaking the GO in water. The nature of equilibrium GO LC has to be determined after GO is completely settled. Due to large aspect ratio of GO, it takes several days for GO to settle down at the bottom of the tubes. Figure 3b shows typical separated and stable nematic phase of GO LCs after 4 h of centrifuge and long-time standing. The volume fraction of nematic phase increases with initial GO mass fraction. Figure 3c shows typical isotropic to nematic (I-N) phase transition, which is broad in general and varies with different GO sources [7]. The broad I-N transition is due to the large size polydispersity (83 %) of GO sheets [52], and the low transition point is due to large aspect ratio of GO [7, 19, 47]. Both observations agree with the Onsager theory model that large aspect ratio (about 2600) means lower phase transition point [53].

As discussed before, GO is negatively charged due to carboxyl and hydroxyl groups on the surface. Like many other liquid crystals, GO LCs will be affected by solvent ion strength and pH level [19, 50]. The effect of phase of GO LC can be seen in Fig. 4a, when more of the NaCl is added, the biphasic GO LC can become isotropic and even collapse resulting in aggregated GO under a high NaCl level. This effect of salt can be understood from the reduced zeta potential shown in Fig. 4b. It is obvious that the repulsive force is dominant in GO dispersions and the dramatic decrease of GO interactions resulted in coagulation of dispersions. A high ion concentration screens the negative charge of GO flakes, leading to reduced electrostatic repulsive force between GO sheets [19, 50, 54].

Tkacz and co-workers investigated GO suspension in water with five different pH levels; the results are shown in Fig. 4c. The I-N transition can only be observed with pH values of 2, 6, and 9 [51]. The pH level will not affect the I-N transition point, but higher pH level will broaden the I-N transition region. This can be attributed to the influence of pH value on the ionization of surface functional groups. Increasing of the pH value improves the charge density of GO flakes and further promotes the fractionation of lateral size. As phase transition of GO dispersion is closely related with flake size/aspect ratio, increasing polydispersity will broaden the biphasic region. For extreme pH levels of 1 and 14, however, GO will form aggregation. In order to create stable GO LCs in solutions with high ionic strength and extreme pH level, Zhao and partners developed an amphiphilic polyelectrolytes with hydrophobic backbone and hydrophilic ionic groups (PHBIG) to absorb on GO sheets by hydrophobic forces from the water [48]. With this PHBIG to reduce the interfacial tension of GO sheets and water, GO LCs can be stably dispersed for relatively long time even in serum. This improvement in maintaining phase stability of GO LCs expands the application range to extreme conditions like electrolyte solutions and biological surroundings.

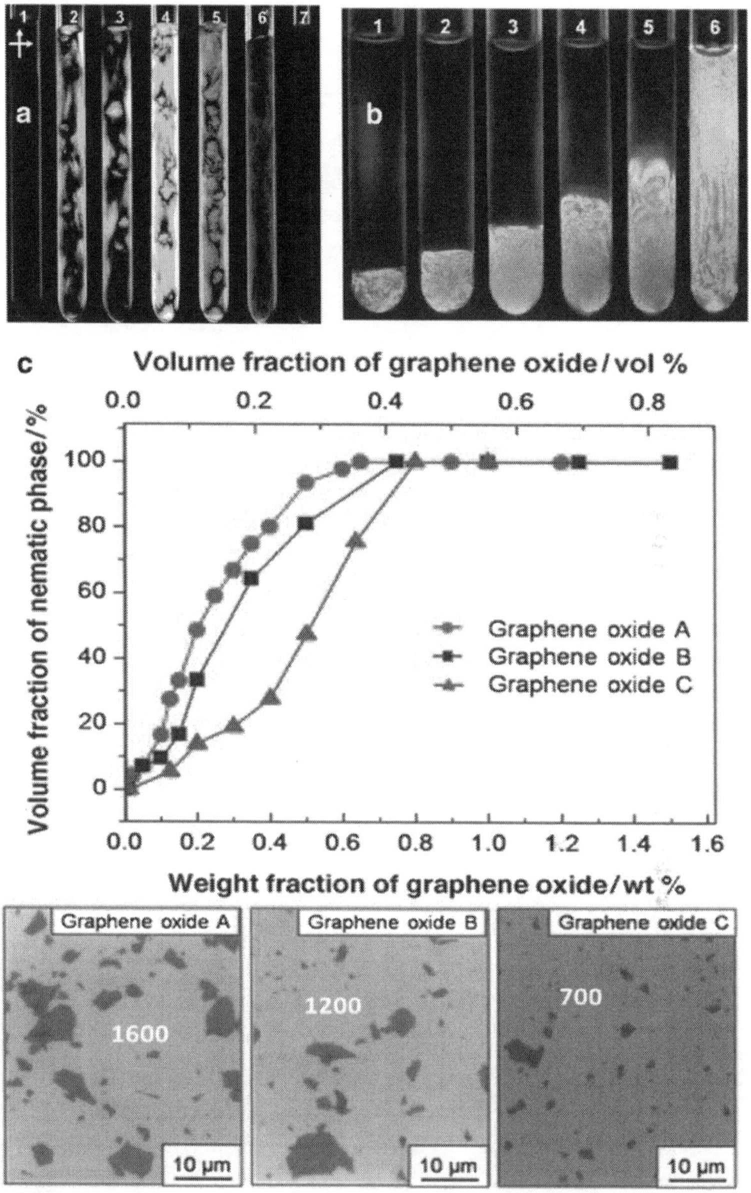

Figure 3 Birefringence images of GO dispersions with various mass fractions. **a** Birefringence images of GO dispersions in test tube with f_m 1×10^{-4}, 2.5×10^{-4}, 5×10^{-4}, 1×10^{-3}, 5×10^{-3}, 1×10^{-2}, and 2×10^{-2} (from 1 to 7) [19]. **b** Birefringence images of GO dispersions when nematic phase is separated from top isotropic phase after centrifuge and long-time standing. No relationship between labels in **a** and **b** [19]. **c** Nematic phase volume fraction versus graphene oxide concentration. Scanning electron microscopy (SEM) images of graphene oxide platelets exfoliated from various graphite sources with D/h aspect ratio of 1600, 1200, and 700, respectively [7]

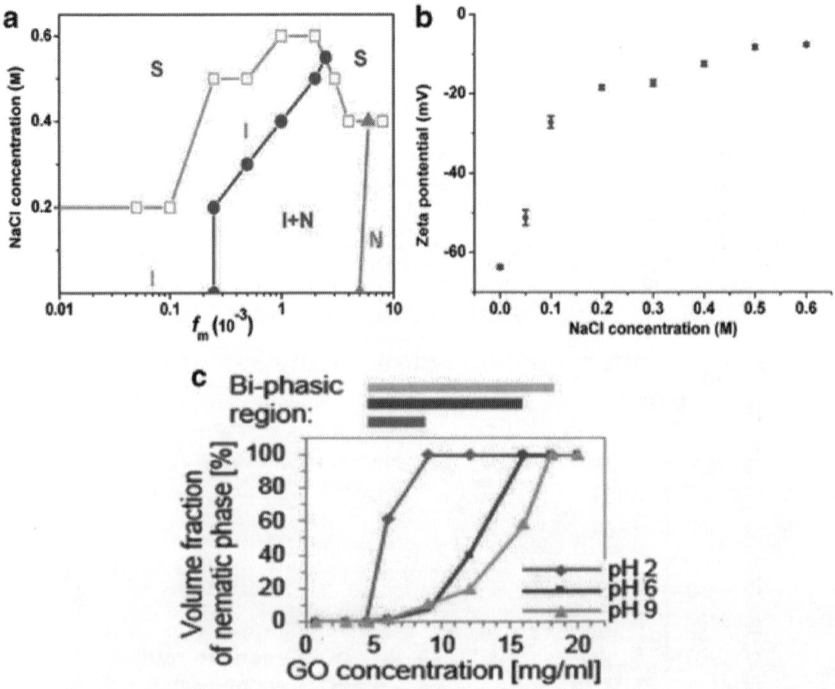

Figure 4 .a Phase diagram of GO based on mass fraction (f_m) and NaCl concentration. "I," "I + N," "N," and "S" represent isotropic phase, isotropic-nematic state, nematic phase, and solid state of GO dispersions. **b** Zeta potential of GO dispersion (f_m = 0.005) versus NaCl concentration [19]. **c** Volume fraction of nematic phase as a function of GO concentration with pH value of 2, 6, and 9 [51]

1.4.2 Rheological Properties

Rheological properties played crucial roles in phase transition and device application of liquid crystals [55, 56]. Xu and Gao observed flow alignment from viscosity decrease during isotropic to nematic phase transition [19]. Then, Yang and partners investigated in detail with scanning electron microscope (SEM) [57]. Figure 5a shows the SEM images of flow alignment of GO dispersions with high and low concentration. The bottom of Fig. 5a shows the schematic of flow-induced alignment of GO LCs. Kumar et al. systematically studied the viscosity versus shear rate of GO suspensions with various shear rate and volume fractions [58]. As shown in Fig. 5b, the viscosity of GO suspensions decreases rapidly to a very small value with increasing of the shear rate. Specifically, GO dispersions exhibit Newtonian behavior in intermediate range at low concentration and typical shear thinning behavior at high concentration. This dramatic decrease of viscosity is resulted from shear alignment of GO sheets [19, 57]. Viscosity does not increase monotonically with GO concentration. As shown in Fig. 5c, at low concentration, the viscosity increases with GO concentration in the isotropic phase. It reaches the maximum and then decreases in the nematic phase with increasing of the volume fraction. The maximum point of viscosity is the transition point from isotropic to nematic phase.

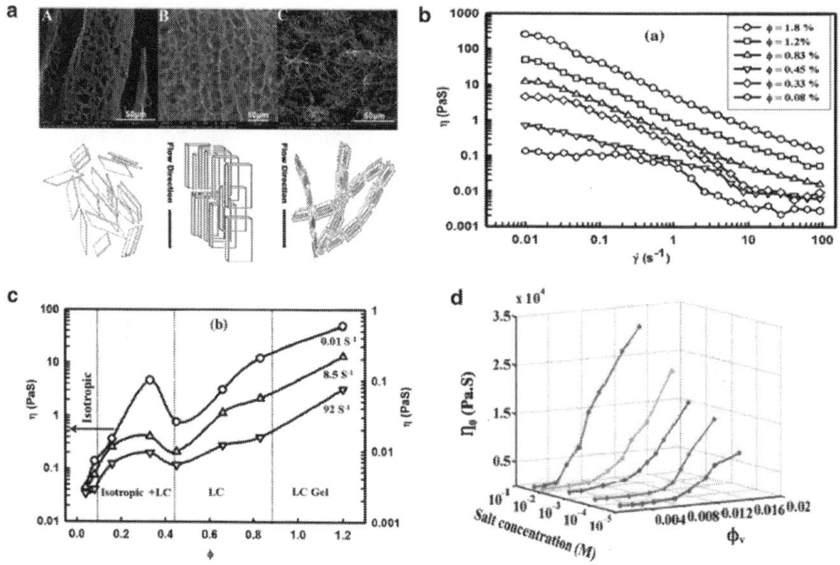

Figure 5. Rheological properties of GO dispersions. **a** Scanning electron microscope (SEM) images of GO dispersion under flow. *A* is the prepared GO dispersion with concentration of 5 mg/ml. *B* is the well-aligned structure under flow of the same GO dispersion. *C* is the flow-induced alignment of low concentration (1 mg/ml) of GO dispersion. *Bottom*: the corresponding flow-induced alignment schematics of GO dispersion [57]. **b** Relations between shear viscosity (η) and shear rate ($\gamma^{\cdot}$) with various volume fraction of GO dispersions. **c** The no-monotonic increasing of viscosity (η) with volume fraction (ϕ) [58]. **d** Zero shear viscosity (η_0) of GO dispersions as a function of volume fraction (ϕ_v) and salt concentration (*M*) [50]

Considering that the addition of salt would affect the phase transition of GO dispersions, Konkena et al. researched the viscosity of GO dispersions based on various salt concentrations [50]. Figure 5d shows the relations between zero shear viscosity and volume fraction as well as salt concentration. The viscosity of GO dispersions increases with volume fraction and salt concentration. At high salt concentration, viscosity increases very fast with the increasing of the volume fraction. However, GO dispersions show relatively small viscosity with low salt concentration while even if the volume fraction is pretty high. The addition of salt reduces the Debye screening length, screening the surface charges and enhancing the surface attractive force [59, 60], therefore, the viscosity goes up with the rising of salt concentration.

Besides viscosity analysis for shear flow and flow-induced alignment, the flow-induced GO flake ordering was measured with polarizing optical method and quantitatively analyzed by Hong et al [61, 62]. In their research, the motion model of GO flakes under flow was built and the corresponding order parameters were calculated.

1.4.3 Magnetic/Electro-Optical Properties

For potential device applications of GO LCs, it is important to be able to manipulate the orientation of GO flakes by external fields or forces [63, 64]. GO exhibits a weak magnetic susceptibility and can be aligned by a magnetic field. The top of Fig. 6a shows the experimental setup, and the birefringence pictures in the bottom figure indicate that the planes of GO flakes are aligned with the magnetic field. The problem is that the alignment process took about several hours because of the very weak magnetism of GO flakes. When GO was functionalized with magnetic iron oxide (Fe_2O_3), the alignment was completed in several seconds.

The alignment of GO by electric field is not straightforward because GO platelets will undergo electrophoretic migration and become electrochemically reduced under DC electric field [7]. Shen and co-workers solved this problem by employing high-frequency alternating current (AC) electric field [65]. The birefringence images in the top of Fig. 6b reveal the alignment of GO sheets with a 10-kHz electric field. The field of 5 Vmm^{-1} is about three orders of magnitude smaller than that for the switching of conventional molecular LCs. One serious problem with GO LCs is that the switching only works in low concentration (0.1 vol.%), not for LCs with higher GO concentration, as shown at the bottom of Fig. 6b.

Ahmad et al. investigated the effect of flake size, electric field, and concentration on field-induced birefringence [66]. Figure 6c shows that birefringence of all four samples increases with the electric field, and GO dispersions with smaller flake sizes exhibit higher birefringence; here, all four samples are in the I-N transition state. Figure 6d shows the dependence of birefringence on GO concentration under the electric field of 20 V/mm. It can be seen that there is an optimal concentration, which is also dependent on the GO flake size. GO LC with a mean size of 0.51 μm has the maximum birefringence at 2 % weight concentration.

The electric field-induced birefringence is a type of Kerr effect observed in nonlinear optical materials. The maximum Kerr coefficient of GO LCs obtained so far is $1.8 \times 10^{-5}\ mV^{-2}$ [65] which is about three orders of magnitude larger than that of other optical materials [64, 67]. This high Kerr coefficient is a consequence of the large anisotropy of the polarizability of GO, relatively large flake spacing [68] as well as electrical double layer from surface oxygen functional groups [68–70]. As with the birefringence, the Kerr effect is dependent on the size and concentration of GO flakes.

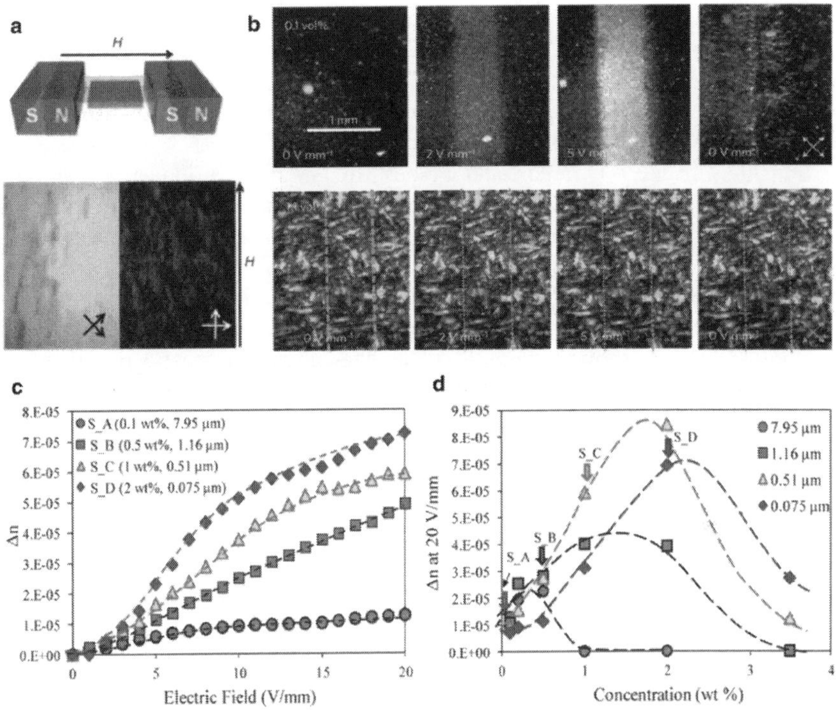

Figure 6. a Alignment of GO LCs with magnetic field. *Top*: experiment diagram of magnetic field-induced alignment. *Bottom*: texture of GO LCs aligned by magnetic field [7]. **b** Birefringence induced by electric field. *Top*: birefringence variation of 0.1 % GO dispersion under 10 kHz and various electric fields. *Bottom*: no birefringence change for 1.1 % GO dispersion with the same electric field [65]. **c, d** GO dispersion birefringence (Δn) as a function of electric field (**c**) and as a function of concentration (**d**) for GO with different sizes under electric of 20 V/mm [66]

Because the ionic strength of solution will affect the interaction between GO flakes, it will certainly modify the birefringence behavior of GO LCs. Figure 7 shows that NaOH has a negligible effect; HCl and NaCl, especially NaCl, can reduce the birefringence by more than a half at a concentration of 10^{-3} M. This is probably because ionic solution is more effective in screening the external field. Therefore, it is important to reduce the concentration of residual salts of oxidation reagents during fabrication of GO dispersions for improved electro-optical performance of GO LCs [71].

1.4.4 Optical Absorption and Fluorescence Properties

Compared with conventional molecular LCs, GO LCs are very unique for strong optical absorption fluorescence in the UV and visible range. Figure 8a shows the absorption spectra of GO and reduced graphene oxide (rGO) [26]. The absorption spectrum of rGO is very similar to that of graphene, consisting of a broad feature extending to infrared wavelength and a UV peak at 275 nm due to π-π* transition from conjugated C-C bonds. In contrast, GO shows very weak absorption at wavelength longer than 500 nm, a shoulder near 300 nm due to n-π* transition from C = O bonds, and a blueshifted π-π* transition at 230 nm [24, 26, 30, 72, 73]. This big change in absorption from graphene to graphene oxide is a result of a change in the chemical composition and associated electronic band structure, namely, a decrease in conjugated C-C bonds and an increase in functional groups during oxidation and exfoliation. These functional groups such as C-O, $-CH_2$, -OH, and -COOH can be observed with Fourier transform infrared (FTIR), as shown in Fig. 8b. The size-dependent spectra reveal that $-CH_2$ is mainly formed on the edge of GO flakes, while other groups are attached to GO surfaces [74].

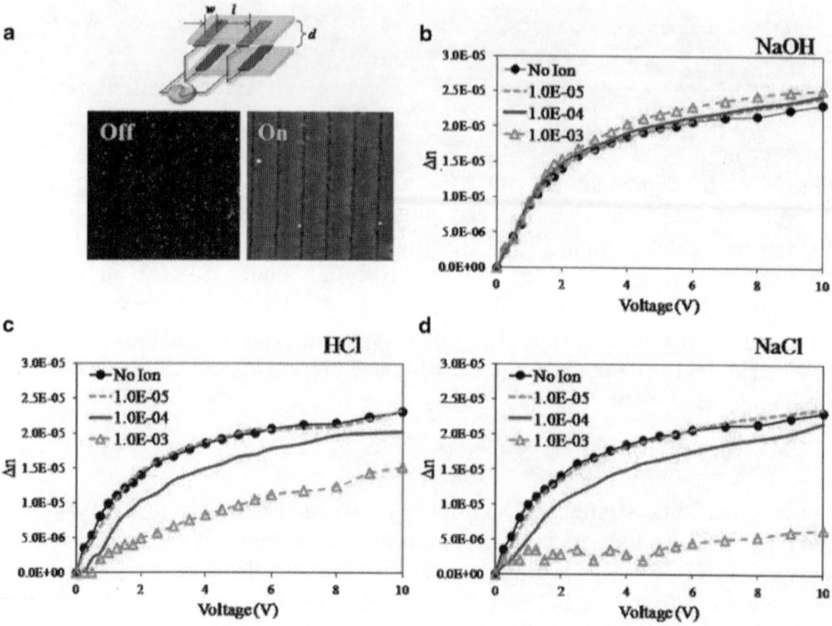

Figure 7. Electro-optical response based on different ionic solutions and concentrations. **a** The test cell model and images with and without electric field. **b–d** Dependence of birefringence (Δ*n*) on electric field and concentration for (**b**) NaOH-GO dispersion, (**c**) HCl-GO dispersion, and (**d**) NaCl-GO dispersion [49]

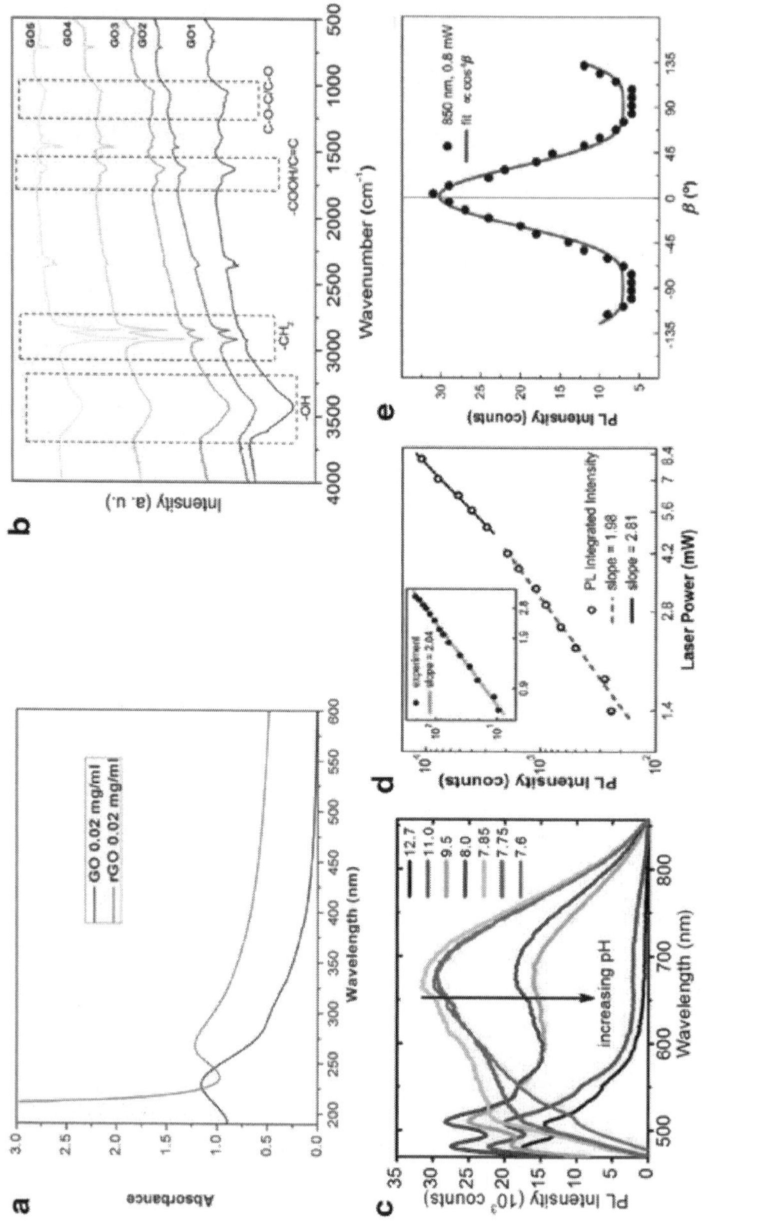

Figure 8. Optical absorption and fluorescence properties of GO. **a** UV-vis absorption spectra of GO dispersion and reduced GO (rGO) [26]. **b** Fourier transform infrared (FTIR) spectra of GO with different sizes. GO1, GO2, GO3, GO4, and GO5 have average size of 390, 200, 140, 60, and 38 nm [74]. **c** Photoluminescence (PL) spectra of GO dispersion with a different pH value [75]. **d** Fluorescence of GO as a function of laser power. **e** The PL intensity as a function of angle β between the plane of GO flake and polarization of excitation light [76]

Unlike graphene, GO exhibits a broadband fluorescence ranging from UV, visible to near-infrared wavelengths [75–81], which opens up new device applications in optoelectronics and display [82]. The photoluminescence (PL) is believed to originate from electro-hole recombination in carbon clusters within carbon-oxygen matrix [79]. The PL of GO can be controlled by pH level [75]. As shown in Fig. 8c, the PL emission is centered around 500 nm in basic condition, but it redshifts to 680 nm in acidic condition. This spectral shift is due to electronic coupling of carboxylic acid groups to atoms of graphene backbone. It was also found that PL intensity was dependent on laser power and polarization of excitation laser [76]. As shown in Fig. 8d, the PL increases quadratically at low incident laser power and grows even faster at higher incident power. This nonlinear dependence of PL intensity can be attributed to multiphoton excitation [76]. The PL is also highly dependent on the angle β between the plane of GO flake and polarization of excitation laser, can be described as $I_{PL} \propto \cos^4 \beta$. As shown in Fig. 8e, PL intensity reaches the minimum when polarization was perpendicular to the flake plane ($\beta = \pm 90°$) and the maximum PL is obtained when polarization was parallel to the flakes ($\beta = 0°$). This property can be used to identify the orientation of GO flakes in GO LCs [76].

1.5 Displays with GO LCs

1.5.1 GO Back-Illuminated Liquid Crystal Display (LCD)

Electric field-induced birefringence is the basis for many device applications of conventional liquid crystals. Similar applications, especially display, are also enabled by large Kerr effect of GO LCs [49, 65, 68]. A prototype of back light-illuminated GO liquid crystal display (LCD) is shown in Fig. 9, where the glass cell is filled with 0.056 vol.% GO LC, and the electric field is 20 V at 10 kHz. Compared with conventional LCDs, the device consumes low power and does not require special treatment for its electrodes [28, 41]. The size of the device, including the spacing between the top and bottom electrodes, can be significantly reduced. One problem with GO LCD is its slow on-off switching time, a few seconds for this device [68].

A natural way to reduce the switching time is to reduce the size of GO flakes [66]. Figure 10a, b shows rising and falling responses of four GO LCs with sizes from ~10 μm down to sub 0.1 μm. The rising and falling time constants are summarized in Fig. 10c. It can be seen that both rising and falling time constants become shorter when the size decreases from 7.95 to 0.51 μm, but the rising time increases when the GO size is further reduced. This difference between rising and falling responses arises from different forces that govern the rotational dynamics of GO flakes. The rising time for the alignment of GO flakes is dependent on the anisotropy of polarizability and rotational viscosity. For very small GO flakes, polarization anisotropy decreases much faster than rotational viscosity, so the rising time increases. In contrast, the falling response is determined by rotational viscosity alone, smaller flakes suffer less from rotational viscosity, leading to reduced falling time. An optimal flake size is around 0.6 μm with average response time about one tenth of a second. Although this time is quite small, shorter switching time is required for certain applications such as TV and computer screens.

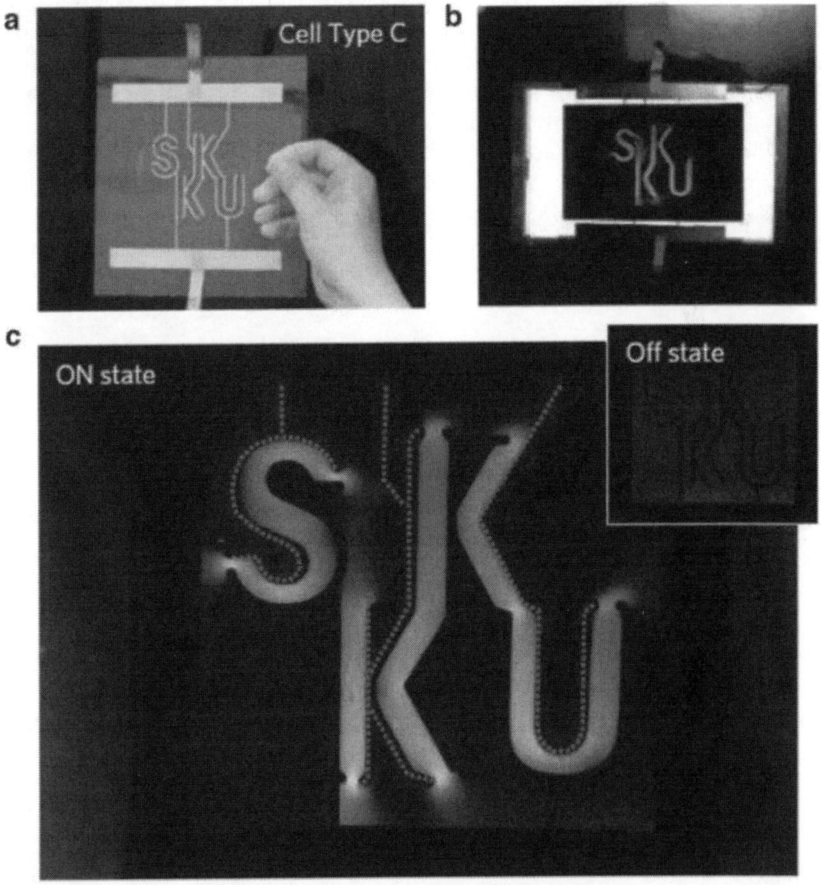

Figure 9 . GO back-illuminated liquid crystal display (LCD) model. **a** Glass substrate with simple wire electrodes. **b** GO LCD on the *top* of back light. **c** Images of the device with electric field on and off [65]

1.5.2 Rewritable and Reflective Displays

In addition to LCD application using birefringence and back-illuminated light, the unique property of GO also enables a new type of display technology. Figure 11 demonstrates that GO LC can be used as a rewritable paper or board. Arbitrary features can be created and erased with a pen or stick [83]. Furthermore, the background of the LC surface can switch between dark as a black board and bright as a white board. This is an excellent example of reflective display that makes use of ambient light and does not need polarizing optics and back illuminating light. Due to low cost and energy consumption, reflective display has been widely used in electronic books such as Kindles. The reflective display in electronic books employ double-colored microcapsules that are black on half surface but white on the other half surface. The orientation of such "black-white pigment" can be controlled by electrical field. The reflective display technique used in Fig. 11 is very different: it makes use of unique property of GO flakes: strong optical anisotropy and absorption. For example, when planes of GO flakes are aligned with LC liquid surface, GO flakes will function as microscale mirrors, producing a shiny and white surface. When GO flakes are randomly oriented or with the planes in perpendicular to LC surface, they appear dark due to weak back scattering and strong absorption. As discussed above that the orientation of GO flakes can be controlled by electrical field, electrically controlled GO-based reflective display is expected in the near future.

2. CONCLUSIONS

GO is unique compared to molecules in conventional LCs in that GO exhibits the largest diameter to thickness aspect ratio and largest shape and optical anisotropy. GO also shows magnetic response and can absorb and emit light in the visible and near-infrared range. These unique properties have made GO LCs very different from conventional ones in terms of basic phase property and device applications. Due to large aspect ratio and non-uniformity in size, stable GO LCs only show nematic phase and a broad phase transition from isotropic to nematic. Besides back-illuminated displays like conventional LCs, GO LCs can also be used in reflective display. The large Kerr effect and birefringence can be used in optoelectronic devices such as spatial phase modulators, Q-switch, saturable absorber, etc. The strong fluorescence allows GO to be used in many optical sensing and new type of display.

There are two big challenges. The first is to produce large quantity of GO with uniform size distribution. Uniform GO will enable new LC phases and will improve performance of GO devices. For example, it can help reduce the response time under electric field. The second challenge is the precise control and alignment of GO with high order parameter. Unlike rod-like molecules in conventional LCs, GO has more orientational degrees of freedom. An electric field cannot completely determine the orientation of a GO flake. A complete control of GO orientation and highly ordered alignment of GO will open up novel device applications.

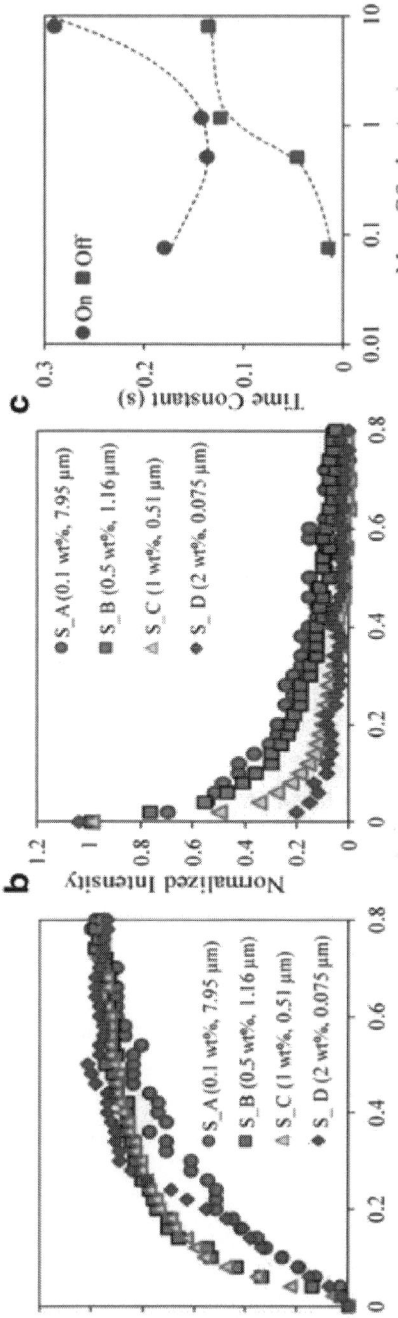

Figure 10. Dynamic response time of GO LCs with different flake sizes. **a**, **b** Rising and falling when electric field is turned on and off. **c** Dynamic response time constants for GO dispersions with various sizes [66]

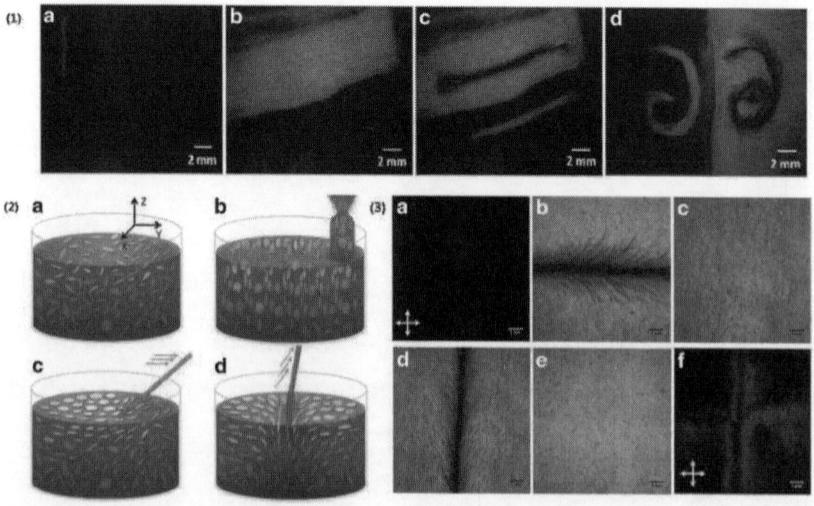

Figure 11. 1 Two display methods on the GO LCs surface. *a* A dark surface with weak light scattering. *b* A bright area created on a dark surface. *c Straight* and (*d*) *curved lines* created on dark and bright surface, respectively. **2** Orientation model of GO LCs during writing and erasing. *a* Isotropic phase of GO dispersions with random orientations. *Z* is perpendicular to the liquid surface. *b* Preparing of dark LC surface injecting GO dispersions with capillary tube onto a petri dish. The directors of GO flakes are parallel to *X-Y* surface. *c* Preparing of a bright surface by sliding a stick along the surface. The directors of GO flakes are perpendicular to LC surface. *d* Preparing of a *dark line* by sliding a stick horizontally in the LCs. **3** The rewritable and erasable properties of GO LCs. *a* Transmission image of isotropic GO LCs between two crossed optic polarizers. *b–e* Reflective images of GO LC surface with procedure of writing-erasing–writing-erasing. *f* Transmission image of GO LCs plotted in *e* [83]

ACKNOWLEDGEMENTS

This work was supported by the Specialized Research Fund for the Doctoral Program of Higher Education of China through SRFDP no. 20120185120037, the National Higher Education Institution General Research and Development Funding through Grant no. ZYGX2012J034, and National Basic Research Program (973) of China through Grant no. 2015CB358600 and 2013CB933801. Jiming Bao acknowledges support from the National Science Foundation (Career Award ECCS-1240510) and the Robert A Welch Foundation (E-1728).

COMPETING INTERESTS

The authors declare that they have no competing interests.

AUTHORS' CONTRIBUTIONS

FL, XT, and ZMW proposed the structure of this review paper. FL wrote the manuscript. JB, YW, and XT improved this manuscript. JB and ZMW revised the final edition of the manuscript. All authors read and approved the final manuscript.

REFERENCES

1. Novoselov KS, Geim AK, Morozov SV, Jiang D, Zhang Y, Dubonos SV et al. (2004) Electric field effect in atomically thin carbon films. Science 306(5696):666–9

2. Geim AK (2009) Graphene: status and prospects. Science 324(5934):1530–4

3. Allen MJ, Tung VC, Kaner RB (2010) Honeycomb carbon: a review of graphene. Chem Rev 110(1):132–45

4. Rao CN, Sood AK, Subrahmanyam KS, Govindaraj A (2009) Graphene: the new two-dimensional nanomaterial. Angew Chem 48(42):7752–77

5. Stankovich S, Dikin DA, Dommett GH, Kohlhaas KM, Zimney EJ, Stach EA et al. (2006) Graphene-based composite materials. Nature 442(7100):282–6

6. Zheng QB, Li ZG, Yang JH, Kim JK (2014) Graphene oxide-based transparent conductive films. Prog Mater Sci 64:200–47

7. Kim JE, Han TH, Lee SH, Kim JY, Ahn CW, Yun JM et al. (2011) Graphene oxide liquid crystals. Angew Chem 50(13):3043–7

8. Brodie BC (1859) On the atomic weight of graphite. Philos Trans R Soc Lond 149:249–59

9. Staudenmaier L (1898) Verfahren zur Darstellung der Graphitsäure. Ber Dtsch Chem Ges 31(2):1481–7

10. Hofmann U, König E (1937) Untersuchungen Über Graphitoxyd. Z Anorg Allg Chem 234(4):311–36

11. Hummers WS, Offeman RE (1958) Preparation of graphitic oxide. J Am Chem Soc 80(6):1339

12. Pei S, Cheng H-M (2012) The reduction of graphene oxide. Carbon 50(9):3210–28

13. Zhu Y, Murali S, Cai W, Li X, Suk JW, Potts JR et al (2010) Graphene and graphene oxide: synthesis, properties, and applications. Adv Mater 22(35):3906–24

14. Loh KP, Bao Q, Eda G, Chhowalla M (2010) Graphene oxide as a chemically tunable platform for optical applications. Nat Chem 2(12):1015–24

15. Dreyer DR, Park S, Bielawski CW, Ruoff RS (2010) The chemistry of graphene oxide. Chem Soc Rev 39(1):228–40

16. Behabtu N, Lomeda JR, Green MJ, Higginbotham AL, Sinitskii A, Kosynkin DV et al (2010) Spontaneous high-concentration dispersions and liquid crystals of graphene. Nat Nanotechnol 5(6):406–11

17. Davis VA, Parra-Vasquez AN, Green MJ, Rai PK, Behabtu N, Prieto V et al. (2009) True solutions of single-walled carbon nanotubes for assembly into macroscopic materials. Nat Nanotechnol 4(12):830–4

18. Ericson LM, Fan H, Peng H, Davis VA, Zhou W, Sulpizio J et al. (2004) Macroscopic, neat, single-walled carbon nanotube fibers. Science 305(5689):1447–50

19. Xu Z, Gao C (2011) Aqueous liquid crystals of graphene oxide. ACS Nano 5(4):2908–15

20. Paredes JI, Villar-Rodil S, Martinez-Alonso A, Tascon JM (2008) Graphene oxide dispersions in organic solvents. Langmuir 24(19):10560–4

21. Stankovich S, Piner RD, Chen X, Wu N, Nguyen ST, Ruoff RS (2006) Stable aqueous dispersions of graphitic nanoplatelets via the reduction of exfoliated graphite oxide in the presence of poly(sodium 4-styrenesulfonate). J Mater Chem 16(2):155–8

22. Stankovich S, Piner RD, Nguyen ST, Ruoff RS (2006) Synthesis and exfoliation of isocyanate-treated graphene oxide nanoplatelets. Carbon 44(15):3342–7

23. Park S, Ruoff RS (2009) Chemical methods for the production of graphenes. Nat Nanotechnol 4(4):217–24

24. Marcano DC, Kosynkin DV, Berlin JM, Sinitskii A, Sun Z, Slesarev A et al. (2010) Improved synthesis of graphene oxide. ACS Nano 4(8):4806–14

25. Peng L, Xu Z, Liu Z, Wei Y, Sun H, Li Z et al. (2015) An iron-based green approach to 1-h production of single-layer graphene oxide. Nat Commun 6:5716

26. Aboutalebi SH, Gudarzi MM, Zheng QB, Kim J-K (2011) Spontaneous formation of liquid crystals in ultralarge graphene oxide dispersions. Adv Funct Mater 21(15):2978–88

27. Zhu Y, Stoller MD, Cai W, Velamakanni A, Piner RD, Chen D et al. (2010) Exfoliation of graphite oxide in propylene carbonate and thermal reduction of the resulting graphene oxide platelets. ACS Nano 4(2):1227–33

28. Eda G, Chhowalla M (2009) Graphene-based composite thin films for electronics. Nano Lett 9(2):814–8

29. Cai D, Song M (2007) Preparation of fully exfoliated graphite oxide nanoplatelets in organic solvents. J Mater Chem 17(35):3678

30. Gudarzi MM, Moghadam MHM, Sharif F (2013) Spontaneous exfoliation of graphite oxide in polar aprotic solvents as the route to produce graphene oxide—organic solvents liquid crystals. Carbon 64:403–15

31. Wang S, Ang PK, Wang Z, Tang AL, Thong JT, Loh KP (2010) High mobility, printable, and solution-processed graphene electronics. Nano Lett 10(1):92–8

32. Dong X, Su CY, Zhang W, Zhao J, Ling Q, Huang W et al. (2010) Ultra-large single-layer graphene obtained from solution chemical reduction and its electrical properties. Phys Chem Chem Phys 12(9):2164–9

33. Jalili R, Aboutalebi SH, Esrafilzadeh D, Konstantinov K, Moulton SE, Razal JM et al. (2013) Organic solvent-based graphene oxide liquid crystals: a facile route toward the next generation of self-assembled layer-by-layer multifunctional 3D architectures. ACS Nano 7(5):3981–90

34. Ogino I, Yokoyama Y, Iwamura S, Mukai SR (2014) Exfoliation of graphite oxide in water without sonication: bridging length scales from nanosheets to macroscopic materials. Chem Mater 26(10):3334–9.

35. Kim DW, Kim D, Min BH, Lee H, Jung H-T. Sonication-free dispersion of large-area graphene oxide sheets using internal pressure from release of intercalated carbon dioxide. Carbon. 2015;88:126–32.

36. Ogino I, Yokoyama Y, Mukai SR (2015) Sonication-free exfoliation of graphite oxide via rapid phase change of water. Top Catal 58(7-9):522–8

37. Sierra U, Álvarez P, Santamaría R, Granda M, Blanco C, Menéndez R (2014) A multi-step exfoliation approach to maintain the lateral size of graphene oxide sheets. Carbon 80:830–2

38. Lee KE, Kim JE, Maiti UN, Lim J, Hwang JO, Shim J et al. (2014) Liquid crystal size selection of large-size graphene oxide for size-dependent N-doping and oxygen reduction catalysis. ACS Nano 8(9):9073–80

39. Wang X, Bai H, Shi G (2011) Size fractionation of graphene oxide sheets by pH-assisted selective sedimentation. J Am Chem Soc 133(16):6338–42

40. Sun X, Luo D, Liu J, Evans DG (2010) Monodisperse chemically modified graphene obtained by density gradient ultracentrifugal rate separation. ACS Nano 4(6):3381–9

41. Botas C, Pérez-Mas AM, Álvarez P, Santamaría R, Granda M, Blanco C et al. (2013) Optimization of the size and yield of graphene oxide sheets in the exfoliation step. Carbon 63:576–8

42. Jia J, Kan C-M, Lin X, Shen X, Kim J-K (2014) Effects of processing and material parameters on synthesis of monolayer ultralarge graphene oxide sheets. Carbon 77:244–54

43. Pan S, Aksay IA (2011) Factors controlling the size of graphene oxide sheets produced via the graphite oxide route. ACS Nano 5(5):4073–83

44. Zhang L, Liang J, Huang Y, Ma Y, Wang Y, Chen Y (2009) Size-controlled synthesis of graphene oxide sheets on a large scale using chemical exfoliation. Carbon 47(14):3365–8

45. Qi X, Zhou T, Deng S, Zong G, Yao X, Fu Q (2013) Size-specified graphene oxide sheets: ultrasonication assisted preparation and characterization. J Mater Sci 49(4):1785–93

46. Zhao J, Pei S, Ren W, Gao L, Cheng HM (2010) Efficient preparation of large-area graphene oxide sheets for transparent conductive films. ACS Nano 4(9):5245–52

47. Dan B, Behabtu N, Martinez A, Evans JS, Kosynkin DV, Tour JM et al. (2011) Liquid crystals of aqueous, giant graphene oxide flakes. Soft Matter 7(23):11154–9

48. Zhao X, Xu Z, Xie Y, Zheng B, Kou L, Gao C (2014) Polyelectrolyte-stabilized graphene oxide liquid crystals against salt, pH, and serum. Langmuir 30(13):3715–22

49. Hong SH, Shen TZ, Song JK (2014) Electro-optical characteristics of aqueous graphene oxide dispersion depending on ion concentration. J Phys Chem C 118(45):26304–12

50. Konkena B, Vasudevan S (2014) Glass, gel, and liquid crystals: arrested states of graphene oxide aqueous dispersions. J Phys Chem C 118(37):21706–13

51. Tkacz R, Oldenbourg R, Mehta SB, Miansari M, Verma A, Majumder M (2014) pH dependent isotropic to nematic phase transitions in graphene oxide dispersions reveal droplet liquid crystalline phases. Chem Commun 50(50):6668–71

52. Bates MA, Frenkel D (1999) Nematic–isotropic transition in polydisperse systems of infinitely thin hard platelets. J Chem Phys 110(13):6553

53. Forsyth PA, Marcelja S, Mitchell DJ, Ninham BW (1977) Onsager transition in hard plate fluid. J Chem Soc Faraday Transact 2 73(1):84

54. Delhorme M, Labbez C, Jonsson B (2012) Liquid crystal phases in suspensions of charged plate-like particles. J Phys Chem Lett 3(10):1315–20

55. Kharchenko SB, Douglas JF, Obrzut J, Grulke EA, Migler KB (2004) Flow-induced properties of nanotube-filled polymer materials. Nat Mater 3(8):564–8

56. Tan Y, Song Y, Zheng Q (2012) Hydrogen bonding-driven rheological modulation of chemically reduced graphene oxide/poly(vinyl alcohol) suspensions and its application in electrospinning. Nanoscale 4(22):6997–7005

57. Yang X, Guo C, Ji L, Li Y, Tu Y (2013) Liquid crystalline and shear-induced properties of an aqueous solution of graphene oxide sheets. Langmuir 29(25):8103–7

58. Kumar P, Maiti UN, Lee KE, Kim SO (2014) Rheological properties of graphene oxide liquid crystal. Carbon 80:453–61

59. Derjaguin B, Landau L. Theory of the stability of strongly charged lyophobic sols and of the adhesion of strongly charged particles in solutions of electrolytes. Progress in Surface Science. 1993;43(1-4):30–59.

60. Verwey EJW. Theory of the Stability of Lyophobic Colloids. The Journal of Physical and Colloid Chemistry. 1947;51(3):631–6.

61. Hong S-H, Shen T-Z, Song J-K (2015) Flow-induced alignment of disk-like graphene oxide particles in isotropic and biphasic colloids. Mol Cryst Liq Cryst 610(1):68–76

62. Hong S-H, Shen T-Z, Song J-K (2014) Flow-induced ordering of particles and flow velocity profile transition in a tube flow of graphene oxide dispersions. Liq Cryst 42(2):261–9

63. Han TH, Kim J, Park JS, Park CB, Ihee H, Kim SO (2007) Liquid crystalline peptide nanowires. Adv Mater 19(22):3924–7

64. Haseba Y, Kikuchi H, Nagamura T, Kajiyama T (2005) Large electro-optic Kerr effect in nanostructured chiral liquid-crystal composites over a wide temperature range. Adv Mater 17(19):2311–5

65. Shen TZ, Hong SH, Song JK (2014) Electro-optical switching of graphene oxide liquid crystals with an extremely large Kerr coefficient. Nat Mater 13(4):394–9

66. Ahmad RT, Hong SH, Shen TZ, Song JK (2015) Optimization of particle size for high birefringence and fast switching time in electro-optical switching of graphene oxide dispersions. Opt Express 23(4):4435–40

67. Hisakado Y, Kikuchi H, Nagamura T, Kajiyama T (2005) Large electro-optic Kerr effect in polymer-stabilized liquid-crystalline blue phases. Adv Mater 17(1):96–8

68. Kim JY, Kim SO (2014) Liquid crystals: electric fields line up graphene oxide. Nat Mater 13(4):325–6

69. Dozov I, Paineau E, Davidson P, Antonova K, Baravian C, Bihannic I et al. (2011) Electric-field-induced perfect anti-nematic order in isotropic aqueous suspensions of a natural beidellite clay. J Phys Chem B 115(24):7751–65

70. Dimiev AM, Alemany LB, Tour JM (2013) Graphene oxide. Origin of acidity, its instability in water, and a new dynamic structural model. ACS Nano 7(1):576–88

71. Shen TZ, Hong SH, Song JK. Effect of centrifugal cleaning on the electro-optic response in the preparation of aqueous graphene-oxide dispersions. Carbon. 2014;80:560–4.

72. Schniepp HC, Li J-L, McAllister MJ, Sai H, Herrera-Alonso M, Adamson DH et al. Functionalized Single Graphene Sheets Derived from Splitting Graphite Oxide. The Journal of Physical Chemistry B. 2006;110(17):8535–9.

73. Dimiev A, Kosynkin DV, Alemany LB, Chaguine P, Tour JM (2012) Pristine graphite oxide. J Am Chem Soc 134(5):2815–22

74. Goncalves G, Vila M, Bdikin I, de Andres A, Emami N, Ferreira RA et al. (2014) Breakdown into nanoscale of graphene oxide: confined hot spot atomic reduction and fragmentation. Sci Rep 4:6735

75. Galande C, Mohite AD, Naumov AV, Gao W, Ci L, Ajayan A et al. Quasi-molecular fluorescence from graphene oxide. Scientific reports. 2011;1:85.

76. Senyuk B, Behabtu N, Pacheco BG, Lee T, Ceriotti G, Tour JM et al. Nonlinear Photoluminescence Imaging of Isotropic and Liquid Crystalline Dispersions of Graphene Oxide. ACS nano. 2012;6(9):8060–6.

77. Sun X, Liu Z, Welsher K, Robinson JT, Goodwin A, Zaric S et al. (2008) Nano-graphene oxide for cellular imaging and drug delivery. Nano Res 1(3):203–12

78. Luo Z, Vora PM, Mele EJ, Johnson ATC, Kikkawa JM (2009) Photoluminescence and band gap modulation in graphene oxide. Appl Phys Lett 94(11):111909

79. Eda G, Lin YY, Mattevi C, Yamaguchi H, Chen HA, Chen IS et al (2010) Blue photoluminescence from chemically derived graphene oxide. Adv Mater 22(4):505–9

80. Viet Cuong T, Hung Pham V, Woo Shin E, Suk Chung J, Hyun Hur S, Jung Kim E et al. (2011) Temperature-dependent photoluminescence from chemically and thermally reduced graphene oxide. Appl Phys Lett 99(4):041905

81. Zhang W, Zou X, Li H, Hou J, Zhao J, Lan J et al. (2015) Size fractionation of graphene oxide sheets by the polar solvent-selective natural deposition method. RSC Adv 5(1):146–52

82. Sheats JR, Antoniadis H, Hueschen M, Leonard W, Miller J, Moon R et al. (1996) Organic electroluminescent devices. Science 273(5277):884–8

83. He L, Ye J, Shuai M, Zhu Z, Zhou X, Wang Y et al. (2015) Graphene oxide liquid crystals for reflective displays without polarizing optics. Nanoscale 7(5):1616–22

CHAPTER 7

Characteristics of Indium Tin Oxide (ITO) Nanoparticles Recovered by Lift-off Method from TFT-LCD Panel Scraps

Dongchul Choi [1], Sung-Jei Hong [2,*] and Yongkeun Son [1,*]

[1] Department of Chemistry and BK21 PLUS School of HRD Center for Creative Convergence Chemical Science, Sungkyunkwan University, Suwon 440-746, Korea

[2] Display Components and Materials Research Center, Korea Electronics Technology Institute, Seongnam 463-816, Korea

ABSTRACT

In this study, indium-tin-oxide (ITO) nanoparticles were simply recovered from the thin film transistor-liquid crystal display (TFT-LCD) panel scraps by means of lift-off method. This can be done by dissolving color filter (CF) layer which is located between ITO layer and glass substrate. In this way the ITO layer was easily lifted off the glass substrate of the panel scrap without panel crushing. Over 90% of the ITO on the TFT-LCD panel was recovered by using this method. After separating, the ITO was obtained as particle form and their characteristics were investigated. The recovered product appeared as aggregates of particles less than 100 nm in size. The weight ratio of In/Sn is very close to 91/9. XRD analysis showed that the ITO nanoparticles have well crystallized structures with (222) preferred orientation even after recovery. The method described in this paper could be applied to the industrial recovery business for large size LCD scraps from TV easily without crushing the glass substrate.

KEYWORDS

recovery; indium-tin-oxide (ITO); nanoparticle; thin film transistor-liquid crystal display (TFT-LCD) panel; scrap

1. INTRODUCTION

Thin film transistor-liquid crystal display (TFT-LCD) has been used as a most popular flat panel display. As the TFT-LCD market grows, significant amount of rare material of indium (In) has been continuously consumed in the display manufacturing industry. Indium has been widely used for transparent conductive oxides (TCO) in the form of indium tin oxide (ITO), because it has unique high optical transparency and electrical conductivity. The In resource is quite limited, e.g., the world's reserve is merely 16,000 tons while the current annual consumption is about 1400 tons [1]. Consequently, global In resources are being exhausted due to their fast consumption rate in information displays, solar cells and lighting [2,3]. Therefore, recycling of indium has become a very important issue these days. A lot of research works aimed at the reuse of In materials has been performed and, as an alternative, the recycling of In from old ITO targets has been studied [4]. However, only a few works aimed at the extraction of In from TFT-LCD panel scraps has been performed. Moreover, we could not find any result on recovery of ITO with the optimal composition ratio of In/Sn as 91/9, which is the characteristics composition of the TCO. Several works reported that In can be recovered by crushing TFT-LCD panel scraps followed by selective extraction. Acquisition of ITO from the TFT-LCD scrap with the optimal ratio is very valuable for cost saving by reducing the number of steps in the recovery process. In this study, a new recycling technology was developed to recover ITO nanoparticles from TFT-LCD scraps. It is called lift-off method because ITO layer is simply lifted off the glass substrate of panel scraps by dissolving color filter (CF) layer which is located between them as in Figure 1. Characteristics such as particle size, thermal property, crystal structure, composition ratio, and purity of the recovered ITO product were investigated.

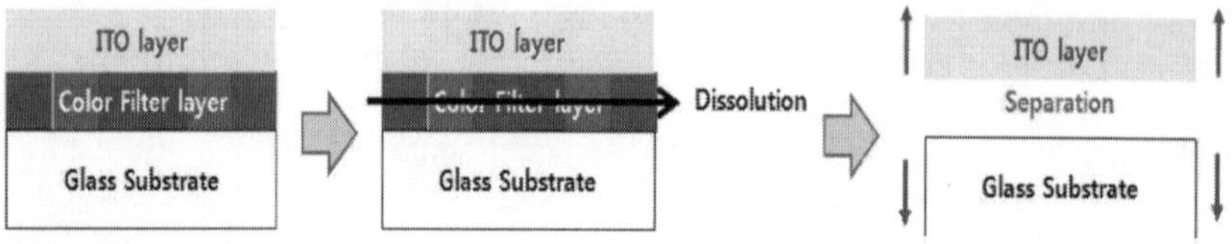

Figure 1. Lift-off method for recovering indium-tin-oxide (ITO) as nanoparticles from glass substrate of thin film transistor-liquid crystal display (TFT-LCD) scraps.

2. EXPERIMENTAL SECTION

An alkaline solution was received from D. Company (Seoul, Korea) and used as a dissolving agent for the color filter. The solution consists of mainly KOH and NaOH and its pH is about 13.4. Polycarbonate (PC) membrane (Iso pore) having 0.4 μm pore diameter was purchased from Milllipore and used as a filter. Methylene Chloride (extra pure grade, Daejung Chemical and Metals, Korea) was used to dissolve the PC membrane. Deionized (DI) water was used to wash out the sample after each step. ITO was taken from a 7 inch size TFT-LCD panel scraps. Disassembly of the TFT-LCDs was done according to previously published methods [5]. First, the upper and lower portions were separated by cutting the edge of the TFT-LCD panel scraps. After cutting, the two plates were easily separated by removing residual liquid crystal with acetone. The upper plate was dipped into the alkaline solution at 80 °C for 30 min. Then the ITO was separated from the glass substrate automatically and remained in the solution after taking the glass substrate out of the solution. The ITO precipitate in the solution was filtered and washed several times with D.I. water. Filter cake was dried in a convection oven at 60 °C for 1 h. Then the thermal behavior of the filter cake was analyzed by thermogravimetric analysis (TGA) and differential thermal analysis (DTA) were performed by using thermogravimetry/differential thermal analyzer (TG/DTA: Seiko Exstar 6000, Seico Inst., Japan). In this way, the heat-treatment temperature to burn out the organic components was determined. The particle size, crystal structure, composition ratio, and purity of the ITO was analyzed by means of high-resolution transmission electron microscope (HRTEM: 300 kV, JEOL, Japan), field-emission scanning microscope (FESEM: JSM-7000F, JEOL, Japan), X-ray diffractometer (XRD: Rigaku Rotaflex D/MAX System) with monochromatic Cu target (λ = 0.1541 nm), inductively coupled plasma (ICP, ICP Optima 43DV, Perkin Elmer, Waltham, MA, USA), X-ray photoelectron spectroscopy (XPS: VG Microtech ESCA2000), using Al Kα radiation as an exciting source. All binding energies (BE) were referenced to the C 1s peak at 284.6 eV.

3. RESULTS AND DISCUSSION

Figure 2a shows the cross-sectional SEM image of 7 inch LCD upper plate taken from a car navigator. The thickness of ITO was observed as 170 nm. Figure 2b shows the simple recovery scheme. The upper plate was immersed into the alkaline solution. After 30 min, the colorless alkaline solution was changed to blue color. That means some parts of RGB pigments were dissolved. This solution was filtered with the polycarbonate (PC) membrane (pore size 0.4 μm).

The filter cake appeared as two parts, film shape part and granular part as in Figure 3a. In the part A, many of irregular film pieces were found. According to the energy dispersive spectrometer (EDS) analysis as shown in Figure 3b its major component was indium. Therefore we estimated that these pieces came from ITO. The granular part B was quite different from the part A. The EDS analysis in Figure 3c shows the elemental components of the aggregate part are carbon, nitrogen, oxygen, copper and magnesium etc. indicating this part came from color filter.

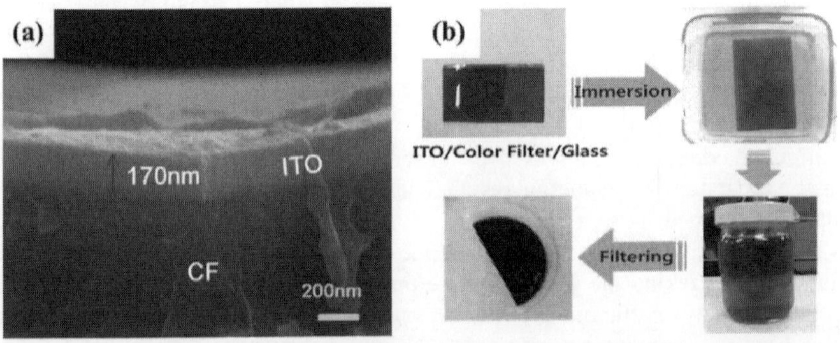

Figure 2. (**a**) SEM image of ITO taken from a car navigator; (**b**) schematic process for the recovery of ITO.

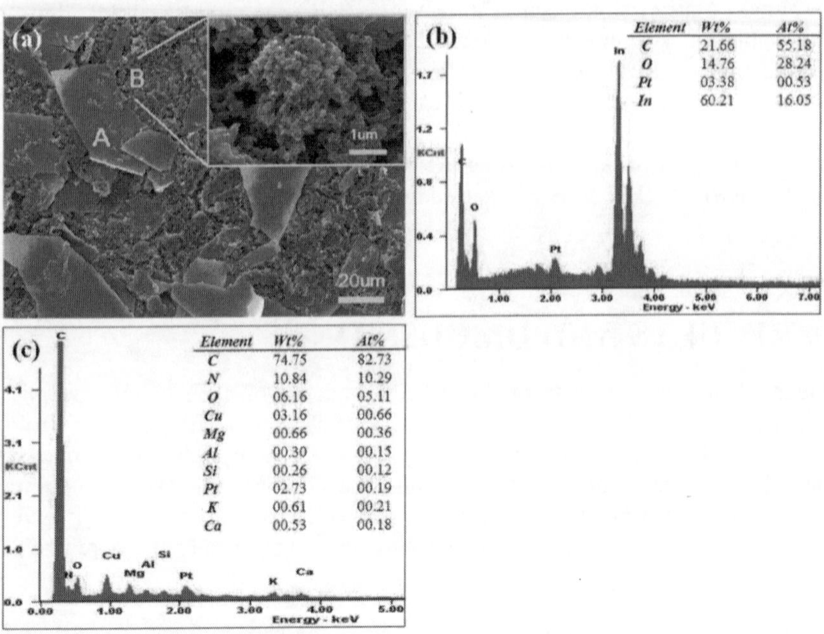

Figure 3. (**a**) SEM image of filter cake; EDS results for the (**b**) part A and (**c**) part B.

The specific analysis of ITO with EDS is very difficult because energy levels of indium and tin are too close to differentiate each other. Therefore, we studied the composition via XPS measurement. The XPS spectra of filter cake are shown in Figure 4. As shown in Figure 4a the peaks located at 444.6, and 452.2 eV corresponding to the In $3d_{5/2}$ and In $3d_{3/2}$ state, respectively. These are related to In^{3+} bonding state in the In_2O_3 [6,7,8,9,10]. Figure 4b shows that Sn $3d_{5/2}$ and Sn $3d_{3/2}$ peaks located at 486.6, 495.2 eV respectively. These binding energies indicate that the Sn is in the Sn^{4+} bonding state from SnO_2 [7,10]. According to the XPS analysis, the binding energy of filter cake, part A, was in good agreement with a typical ITO. Therefore, it is concluded that the recovered sample was mostly consisted of ITO.

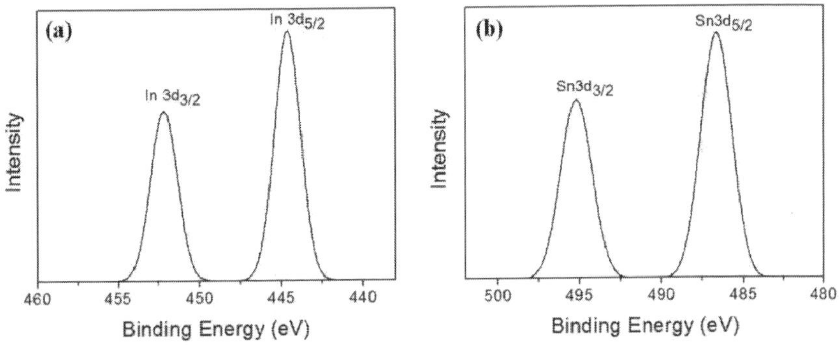

Figure 4. XPS spectra of filter cake part A; (**a**) Indium 3d and (**b**) Tin 3d.

Most of the filter cake was collected by dissolving the filter with the filter cake with methylene chloride. After decanting the methylene chloride away, the residue was washed with same solvent several times. Thermal behavior of the residue was performed by using TG/DTA. As shown in Figure 5a, TG/DTA curves were obtained in the range of ambient temperature to 1000 °C at a heating rate of 10 $°C \cdot min^{-1}$ in air at atmospheric pressure. The first weight loss was observed in temperature range, i.e., between 200 °C and 400 °C, and DTA curve shows the exothermic reaction showing a peak at around 400 °C. This weight loss maybe due to decomposition of CF pigments in the residue. Composition of the CF materials really depends on the producing company but most of them are based on anthraquinone (red) and phtalocyanine (green or blue) [11]. These materials show the weight loss and exothermic reaction at temperatures lower than 600 °C [12,13,14]. These results indicate that thermal decomposition of the CF pigments could occur at lower than 600 °C. The inset in Figure 5a shows the yellow ITO powder obtained after heat treatment at 600 °C for 1 h in air. The color was similar to that of commercial ITO powder. The morphology of recovered ITO was observed by using HRTEM. The sizes of recovered ITO particles were less than 100 nm in diameter and appeared as aggregate as shown in Figure 5b.

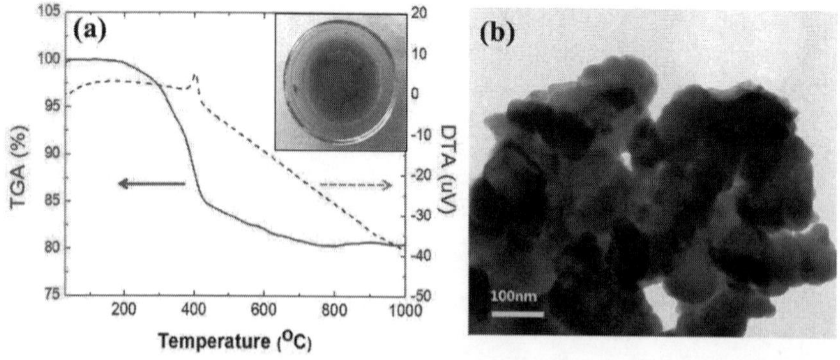

Figure 5. Thermal behavior (**a**) and high-resolution transmission electron microscope (HRTEM) image (**b**) of the recovered ITO powder from TFT-LCD panel after heat treatment at 600 °C.

Figure 6 shows the X-ray diffraction pattern of the recovered ITO powder after heat-treatment at 600 °C. This pattern indicates that our sample has quite similar structure compared to a typical ITO crystal structure. The most intensive diffraction peak appeared at 30.5° and other peaks were at 35.5°, 50.9° and 60.7°. This pattern matches well with the diffraction pattern for the (222), (400), (440) and (622) orientations of cubic crystalline ITO [15,16]. Thus, the ITO particles after heat-treatment at 600 °C are well crystallized as cubic structure. However, the splitting of diffraction peak appeared at 30° indicates that the recovered ITO consists of two different layers. According to the reported results [17,18,19], splitting of XRD peak in ITO films was attributed to existence of two differently strained layers, i.e., amorphous and crystalline layer. The diffraction pattern of the recovered material was similar to that of pure ITO. It is supposed that most In atoms are uniformly substituted to Sn atoms in the lattice [20].

The composition ratios of In and Sn was evaluated before and after heat treatment by using ICP analysis. In and Sn compositions before and after heat treatment were summarized in Table 1. The composition of In and Sn before heat treatment was 91.2 wt.%, 8.8 wt.%, but after heat treatment it was changed to 91.6 wt.%, 8.4 wt.%. The compositions were changed a little after heat treatment. Unfortunately, the reason is not clear up to now and is being under investigation. Nevertheless, the composition was exactly closed to 91/9 even after heat treatment at higher temperatures.

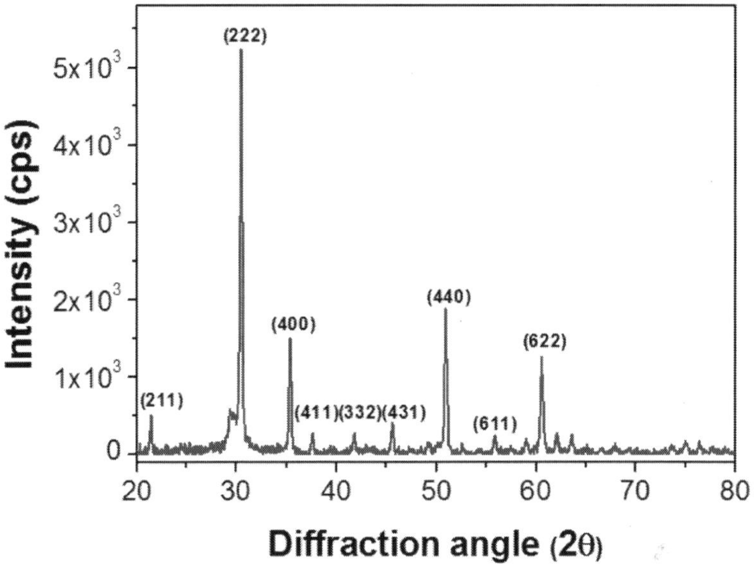

Figure 6. XRD pattern of ITO nanoparticles recovered after heat-treatment at 600 °C.

Table 1. Composition ratios of ITO nanoparticles obtained from TFT-LCD scraps by ICP analysis.

Elements	Before heat treatment (wt.%)	After 600 °C heat treatment (wt.%)
In	91.2	91.6
Sn	8.8	8.4

4. CONCLUSIONS

We have successfully recovered ITO in the form of nanoparticle from TFT-LCD panel scraps by using a simple lift-off method. This lift-off approach provided us a useful recovery method with high yield of 90%. ITO was easily separated as nanoparticles from glass substrate of TFT-LCD scarp. The size of the ITO nanoparticle was less than 100 nm, and appeared as aggregates. The recovered ITO was well crystallized to a (222) preferred orientation, and the composition ratio of In to Sn of about 91 to 9. The lift-off method described in this paper could be easily applied to the large scale recovery business dealing large size LCD panels from TV and other displays.

ACKNOWLEDGMENTS

This work was supported the Energy and Resource Recycling of the Korea Institute of Energy Technology Evaluation and Planning (KETEP) grant funded by the Korea government Ministry of Trade, Industry and Energy (No. 2010501010002B).

AUTHOR CONTRIBUTIONS

Dongchul Choi contributes as a 1st author. Sung-Jei Hong and Yongkeun Son equally contributes as corresponding author.

CONFLICTS OF INTEREST

The authors declare no conflict of interest.

REFERENCES

1. Global and china ITO sputtering targets industry research report 2013–2016. Available online: http://www.researchmoz.us/global-and-china-ito-sputtering-targets-industry-report-2013-2016-report.html (accessed on 26 November 2014).

2. Minami, T. Substitution of transparent conducting oxide thin films for indium tin oxide transparent electrode applications. Thin Solid Films **2008**, 516, 1314–1321.

3. Polinares. Fact Sheet: Indium; Polinares: Dundee, UK, 2012.

4. Yu, J.-K.; Kang, S.-G.; Jung, K.-C.; Han, J.-S.; Kim, D.-H. Fabrication of nano-sized ITO powder from waste ITO target by spray pyrolysis process. Mater. Trans. **2007**, 48, 249–257.

5. Hong, S.-J.; Kim, M.-S.; Kim, J.-W.; Shin, M. Characteristics of indium-tin-oxide (ITO) glass re-used from old TFT-LCD panel. Mater. Trans. **2012**, 53, 968–972.

6. Fan, J.C.; Goodenough, J.B. X-ray photoemission spectroscopy studies of Sn-doped indium-oxide films. J. Appl. Phys. **1977**, 48, 3524–3531.

7. Kim, J.; Ho, P.; Thomas, D.; Friend, R.; Cacialli, F.; Bao, G.-W.; Li, S. X-ray photoelectron spectroscopy of surface-treated indium-tin oxide thin films. Chem. Phys. Lett. **1999**, 315, 307–312.

8. Zhu, F.; Huan, C.; Zhang, K.; Wee, A. Investigation of annealing effects on indium tin oxide thin films by electron energy loss spectroscopy. Thin Solid Films **2000**, 359, 244–250.

9. Pammi, S.; Chanda, A.; Seong, N.-J.; Yoon, S.-G. Growth of high-quality ITO thin films at low temperature by tuning the oxygen flow rate using the nano-cluster deposition (NCD) technique. Chem. Phys. Lett. **2010**, 490, 234–237.

10. Choi, D.; Kim, Y.S.; Son, Y. Recovery of indium tin oxide (ITO) and glass plate from discarded TFT-LCD panels using an electrochemical method and acid treatment. RSC Adv. **2014**, 4, 50975–50980.

11. Kopacek, B. ReLCD recycling and re-use of LCD panels. In Proceedings of the 19th Waste Management Conference of the IWMSA (WasteCon2008), Durban, South Africa, 6–10 October 2008; pp. 6–10.

12. Park, J.; Park, Y.; Park, J. Synthesis and physical property measurement of new red pigment based on anthraquinone derivatives for color filter pigments. Mol. Cryst. Liq. Cryst. **2011**, 551, 116–122.

13. Shirai, H.; Kobayashi, K.; Takemae, Y.; Suzuki, A.; Hirabaru, O.; Hojo, N. Functional metal-porphyrazine derivatives and their polymers, 2. Synthesis and properties of polyimides containing metal-phthalocyanines. Die Makromol. Chem. **1979**, 180, 2073–2084.

14. Bilgin, A.; Yağcı, Ç.; Yıldız, U.; Özkazanç, E.; Tarcan, E. Synthesis, characterization, aggregation and thermal properties of a novel polymeric metal-free phthalocyanine and its metal complexes. Polyhedron **2009**, 28, 2268–2276. [

15. Kulkarni, A.; Schulz, K.H.; Lim, T.; Khan, M. Dependence of the sheet resistance of indium-tin-oxide thin films on grain size and grain orientation determined from X-ray diffraction techniques. Thin Solid Films **1999**, 345, 273–277.

16. Sugiyama, K.; Ishii, H.; Ouchi, Y.; Seki, K. Dependence of indium-tin-oxide work function on surface cleaning method as studied by ultraviolet and x-ray photoemission spectroscopies. J. Appl. Phys. **2000**, 87, 295–298.

17. Yi, C.H.; Shigesato, Y.; Yasui, I.; Takaki, S. Microstructure of low-resistivity tin-doped indium oxide films deposited at 150–200 °C. Jpn. J. Appl. Phys. **1995**, 34, L244–L247.

18. Adurodija, F.; Izumi, H.; Ishihara, T.; Yoshioka, H.; Motoyama, M. Effects of stress on the structure of indium-tin-oxide thin films grown by pulsed laser deposition. J. Mater. Sci. Mater. Electron. **2001**, 12, 57–61.

19. Izumi, H.; Adurodija, F.O.; Kaneyoshi, T.; Ishihara, T.; Yoshioka, H.; Motoyama, M. Electrical and structural properties of indium tin oxide films prepared by pulsed laser deposition. J. Appl. Phys. **2002**, 91, 1213–1218.

20. Nadaud, N.; Lequeux, N.; Nanot, M.; Jove, J.; Roisnel, T. Structural studies of tin-doped indium oxide (ITO) and $In_4Sn_3O_{12}$. J. Solid State Chem. **1998**, 135, 140–148.

CHAPTER 8

Cellular Automata Simulation of Gap-Filler Dissolution during Transient Liquid Phase Bonding of Single Crystal Materials

Adam Ghoneim, Olanrewaju Akanbi Ojo*

Department of Mechanical and Manufacturing Engineering,
University of Manitoba, Winnipeg, Canada

ABSTRACT

A new numerical model is developed using a Cellular Automata (CA) method to study the liquid-phase dissolution behavior of gap-filler powder particles in interlayer powder mixture during transient liquid phase (TLP) bonding process. The model prediction of microstructural evolution in TLP joint between single crystal substrates show that formation of misoriented stray-grains results from incomplete liquation of the gap-filler powder particles. In contrast to what is generally assumed and reported, numerical calculations coupled with experimental verification show that under properly selected process parameters, complete melting of the gap-filler powder particles is possible. This is imperative to prevent the formation of misoriented stray-grains and maintain single crystallinity during TLP bonding of single crystal materials. The dependence of complete melting of the gap-filler particles on salient TLP bonding parameters are analyzed and discussed.

KEYWORDS

Cellular Automata; Transient Liquid Phase Bonding; Single Crystal; Powder Dissolution

1. INTRODUCTION AND BACKGROUND

Single crystal nickel-base superalloys are new generation heat resistant materials used in the manufacture of section components in aircraft and land-based power-generation turbine engines due to their remarkable mechanical properties at elevated temperatures [1]. Prolonged operations under these severe conditions often lead to degradation due to creep, thermo-mechanical fatigue and oxidation. Due to the high cost of these materials, it is often more economical to repair the damaged components. However, conventional repair techniques applied on precipitation hardened superalloys such as fusionwelding methods are severely limited due to the high susceptibility of these alloys to heat affected zone cracking [2] and the possible formation of stray grains in single crystal and directionally solidified materials [3,4]. Furthermore, conventional brazing methods result in the formation of hard and brittle eutectic-type micro-constituents in brazement, which degrades the mechanical properties of brazed materials [2,5].

Transient liquid phase (TLP) bonding has evolved from conventional brazing as a method to prevent the formation of deleterious non-equilibrium solidification micro-constituents. In TLP bonding, a thin layer of filler alloy, rich in melting-point depressant (MPD) solute, is sandwiched between two solid base materials and the entire assembly is heated to the bonding temperature. At the bonding temperature, the filler alloy melts and rapidly attains equilibrium at the liquid-solid interfaces through base-alloy melt-back dissolution process. Following the dissolution stage, a solid-state diffusion of the melting point depressant (MPD) element away from the liquid interlayer into the solid substrate occurs, which results in a decrease in the volume of liquid that can be maintained at equilibrium, thus, causing isothermal solidification to proceed inward from the solid mating surfaces [5]. Following the isothermal solidification stage, the joint is homogenized either at the bonding temperature or at some lower temperature [2,3]. Qualitatively, the TLP bonding technique relies on the variation of liquidus temperature of an alloy with composition and the fact that concentration variation in an inhomogeneous alloy can cause localized melting when the bulk of the material remains solid. The bonding process is ideal for joining base materials which are inherently susceptible to hot cracking or post-weld heat treatment cracking problems and can even be used for joining of dissimilar materials [2,3,5].

Liquid-phase dissolution of the base-material, a phenomenon also known as "liquid-metal erosion", occurs during the initial stage of TLP bonding, as the interlayer liquid tries to attain equilibrium solute concentration with the solid substrate. Liquid-phase erosion of the basematerial causes increased gap width during bonding. An approach that is commonly used to reduce base-material erosion while enriching the joint region with base-material alloying elements involves the use of composite powder mixture as interlayer material. The composite powder mixture consists of commercial filler alloy that contains MPD element and a gap-filler powder that is essentially free of the MPD element, possibly the basematerial powder. Optimization of the process parameters is essential in obtaining a superior joint [6] since partial melting of the gap-filler particles can occur and the unmelted remaining solid components can lead to misoriented stray-grains formation at the joint. Huang et al. [7], investigated the diffusion bonding of Al-based dissimilar composites by using Al-Si, Al-Cu and Al-Si-SiC powders as fillers. They reported that the segregation of SiC particles and formation of porous zones at the joint lead to a reduction of the joint shear strength. They also reported an increase in the shear strength of the joint with increasing temperature and holding time. Wu et al. [8] also investigated wide gap bonding of stainless steel to Ni-base superalloy. They reported that during the bonding process, the gap-filler particles reacted with the liquid and dissolved into the liquid but were incompletely melted and provided the necessary capillary force to retain the molten filler material, which would otherwise be overly fluid to bridge the gap faying surfaces. Zhuang and Eager [9] experimented with powder TLP bonding by using metallic powders coated with a small amount of material that contained a MPD solute. They reported that the coating material and the thickness of the deposit influence porosity formation within the joint. Two different powder materials, Ni-20Cr and 304L stainless steels, both of which were coated with Ni-10P, to join 304L stainless steel base alloy. They reported joints with good mechanical properties when the Ni-20Cr powder was used, while inferior properties are obtained by using the 304L powder, which was attributed to residual porosity observed within the joint. To reduce porosity, they proposed a liquid infiltration approach, where a thin foil that contained a MPD solute is inserted between the base material and the powder mixture.

The technological merits of the composite powder mixture and its application to joining of single crystal materials is generally avoided due to formation of straygrains within the joint region, which compromises high temperature properties of bonded materials. However, the understanding of the dissolution behaviour of gapfiller powder particles by molten filler alloy, which could influence formation of the stray-grains, is limited due to the large number of process and material parameters that affect the final joint microstructure and performance. Mathematical modeling, which is imperative to adequate understanding of the gap-filler powder dissolution process, is very limited due to certain difficulties. Current analytical models [10-13] are only capable of providing an overall estimate of the time required to achieve complete isothermal solidification and are not suitable for simulating powder dissolution. The large number of powder particles results in a large number of particle-liquid boundaries which need to be effectively and efficiently tracked in order to study changes to the powder particles with time. The topological changes of the dissolving or growing powder particle also pose difficulties that can not be addressed by available 1-D numerical TLP bonding models [14-16]. At any time, a powder particle-liquid interface may break up, or join with neighboring interfaces. This poses severe difficulties when conventional interface-tracking methods are used. Additionally, the dissolution of powder occurs rapidly compared to the isothermal solidification stage. Therefore, different time stepping scales are required to capture the intricate changes to the powder particles.

To study the evolution of interfaces, conventional interface-tracking methods would involve the introduction of nodal points at the interface and solving the general transport equations using moving mesh techniques to ensure conformance of the mesh with the moving interface. While the method has the advantage of representing the interface directly, it is very sensitive to topological variations. Therefore, inaccuracies may arise when nodal points at the interface get too close together or too far apart. As a result, to ensure simulation validity, an extra step is generally required to either explicitly remove or add nodal points, depending on the position of the nodes relative to one another. Moreover, major difficulties can be encountered when an interface breaks apart or when interfaces merge together which requires explicit handling of nodal points at the interface and requires remeshing to conform to the new interfacial geometries. The special attention needed to mitigate these problems is rather cumbersome, and as a result, interface-tracking techniques are often limited to relatively simple geometries and interfacial topologies.

CA methods mitigate some of the issues involved in the conventional interface-tracking methods by discritizing the domain into cells where the evolution of each cell is dependent on the states of its immediate neighboring cells. A cellular automaton is a dynamical system, in which space, time, and the states of the system are discrete. It was first developed by John Von Neumann in 1963 after a suggestion by Ulam [17] as a means to study universal copiers and destructors. It was later analyzed by Wolfram [18] in the 1980's as part of his work on analyzing complexity. CA methods provide a frame work to simultaneously study changes in the entire computational domain. This makes the method very useful in studying microstructural evolution and studying complex systems. It can also provide fast implementation on parallel computers due to the inherent parallelism of the CA structure. CA methods applied in materials science generally involve discritizing the microstructure into cells where each cell can have a finite number of states such as concentration, temperature, or phase order parameter. Transition rules are imposed to determine cell evolution depending on the states of its neighboring cells, which can be a Neumann-type or Moore-type, in a square cell lattice. The basic idea of CA for phase change studies is to mimic interface propagation by capturing transition from nearby liquid cells to the solid body according to certain criteria. It is assumed that three kinds of possible cells exist within a simulation domain i.e. solid, liquid and interface cells. Solid cells have a solid fraction of 1, liquid cells have a value of 0, and interface cells have a solid fraction which varies between 0 and 1. Cell capturing rules can be applied to calculate the future solid fraction in each cell, depending on the solutal and thermal distribution in the domain. In CA, the interface is not explicitly tracked like in interface-tracking methods. However, the location of the interface can be obtained by determining the cells where the solid fraction varies from 0 to 1. As a result, the CA method can capture complex interfacial dynamics with great efficiency and its computational requirements (memory and time) are much lower than other methods such as phase-field methods [19]. A review of CA modeling of microstructural evolution has been presented by Janssens [19] and He et al. [20]. It has been used for modeling structural evolution during solidification [21-23]. The method has been also applied to study microstructural evolution during solidstate recrystallization and growth [24]. CA methods has also been coupled with other numerical methods such as the finite difference method [25], the finite volume method [26], the finite element method [27,28] and more recently the lattice Boltzmann method [29].

In the present work, the CA method is used to study the dissolution of solid powder particles during TLP bonding. The mathematical formulation and implementation will be discussed in the following sections followed by the results and discussion.

2. FORMULATION AND IMPLEMENTATION OF THE CELLULAR AUTOMATA NUMERICAL MODEL

A two-dimensional Euclidean lattice Ω that is discretized into square cells is considered (Figure 1). A cell $z_{i,j}$ in the domain hold a finite set of possible states namely, solid, liquid and interface and has a finite set of Moore-type neighboring cells:

$$\left(Z_{i+1,j}, Z_{i-1,j}, Z_{i,j+1}, Z_{i,j-1}, Z_{i+1,j+1}, Z_{i-1,j-1}, Z_{i-1,j+1}, Z_{i+1,j-1}\right).$$

In this work, each cell $z_{i,j}$ holds a finite number of associated data namely, temperature T, composition C, spin index S, and solid-fraction f. The spin index S represents the crystallographic orientation of a powder particle, which is 0 in liquid and can be any integer number in a solid phase. The solid fraction, f, can vary from 0 to 1 in any given cell. A cell having an f value of 0 is a completely liquid cell, a value of 1 is completely solid, and cells having values $0 < f < 1$ define an interface cell.

For the purposes of our analysis, at any given time, determination of the volume fraction of residual solid is more essential than the explicit location of the liquidpowder particle interfaces. Therefore, the basic idea of the current model is to calculate the evolution of the solid fraction in each cell instead of directly tracking the interface motion between a powder particle and the liquid. This approach gives a good representation of evolution of the powder particles while avoiding having to explicitly track the large number of powder-liquid interfaces and can easily handle topological changes and interfacial dynamics such as splitting and joining.

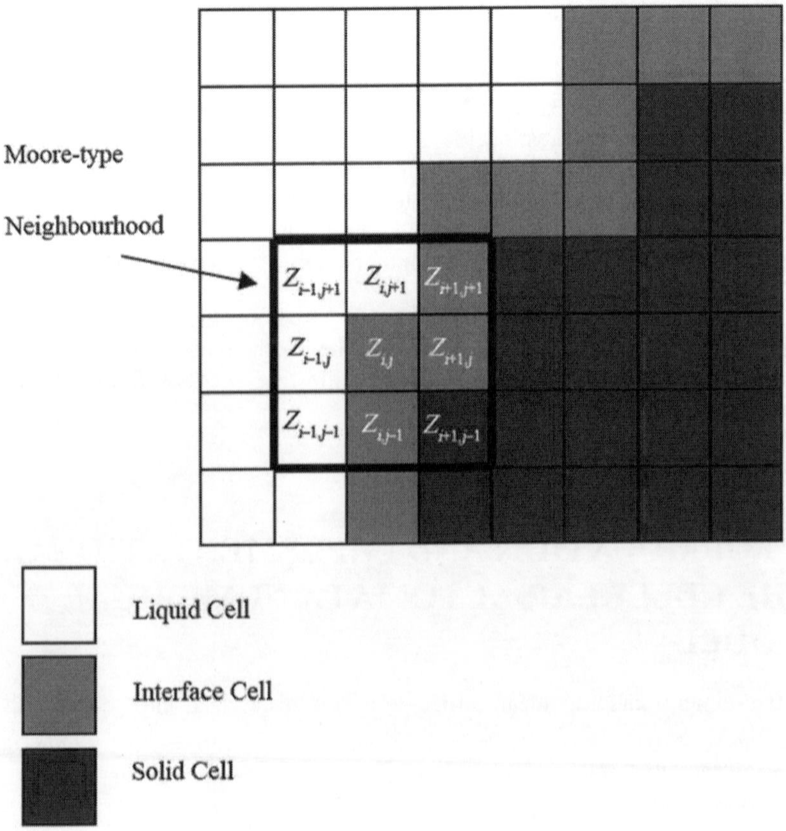

Figure 1. Schematic illustrating the possible cell states and Moore-type neighborhood used in the CA model.

The liquid phase has an initial solute concentration C_F and the solid particles have an initial solute concentration C_m. Solute diffusion in the liquid phase is expressed by:

$$\frac{\partial C_L}{\partial t} = D_L \left(\frac{\partial^2 C}{\partial x^2} + \frac{\partial^2 C}{\partial y^2} \right) \tag{1}$$

Similarly, solute diffusion in the solid particles is expressed as:

$$\frac{\partial C_S}{\partial t} = D_S \left(\frac{\partial^2 C}{\partial x^2} + \frac{\partial^2 C}{\partial y^2} \right) \tag{2}$$

where D_S and D_L are the diffusivities in the solid and liquid cells, respectively. The increase in the solid fraction Δf_S is calculated using [26]:

$$\Delta f_S = \frac{\Delta t}{\Delta X}\left(V_x + V_y - V_x V_y \frac{\Delta t}{\Delta X}\right) \tag{3}$$

where Δt is the time step, ΔX is the cell size, V_x and V_y represent the velocity of the interface cells in the x and y directions. These velocities are calculated using:

$$V_x = \left(\frac{1}{C_L^* - C_S^*}\right)\left\{-D_L \frac{\partial C_L}{\partial x} + D_S \frac{\partial C_S}{\partial x}\right\} \tag{4}$$

$$V_y = \left(\frac{1}{C_L^* - C_S^*}\right)\left\{-D_L \frac{\partial C_L}{\partial y} + D_S \frac{\partial C_S}{\partial y}\right\} \tag{5}$$

where C_L^* and C_S^* are the equilibrium liquidus and solidus concentrations at the interface cells, respectively. At any given time, curvature of the liquid-solid interface is calculated using the procedure proposed by Sasikumar et al. [30], where the number of cells in the extended neighborhood of the interface cell is counted. For example, a 7 × 7 square of cells, which yields a total area of $N_t = 49\Delta X^2$, is drawn around the cell of interest and the number of solid fraction, N_S, in the region is determined. The value for curvature κ can then be found from:

$$\kappa = \frac{1}{\Delta X}\left(1 - \frac{2N_S}{N_t}\right) \tag{6}$$

To calculate the concentrations in the liquid and solid, a typical cell-centered fully implicit transient finite difference method is used to solve Equations (1) and (2). An identical square mesh was used for both the CA and the finite difference computations. Even though a fully implicit transient method is used to solve the concentration equations, there is still a critical time step that must not be exceeded to ensure that the interface velocity is not too high to cause instabilities in the CA solution. The critical time step is calculated from [31]:

$$\Delta t = 0.25 \min\left(\frac{\Delta x}{V_n^{\max}}, \min\left(\frac{\Delta X^2}{D_L}, \frac{\Delta X^2}{D_S}\right)\right) \tag{7}$$

where $V_n^{\max}$ is the maximum velocity of the interface cells. The interface velocities and solute concentrations are calculated under the following assumptions:

1) The molar volume is assumed to be constant in both the liquid and solid. This assumption has been used in previous numerical models [14-16,32] and it yielded good agreement with experimental results.

2) The diffusivity of the MPD solute is independent of concentration. Due to the lack of extensive experimental thermodynamic data on how the localized diffusion of the MPD solute is affected by concentration during TLP bonding, this assumption is commonly used in existing numerical TLP bonding models [14-16,32].

3) There is negligible liquid flow due to convection and stirring in the liquid. In the present experimental work there was no induced convection in the liquid.

In the present model, powder particles are generated and randomly distributed in the domain such that the initial volume fraction of solid in the computational domain is equal to some user specified value. A randomly generated spin index is produced and assigned to each particle to represent its crystallographic orientation. Interface cells are defined and initial solute concentrations in the solid particles, the liquid and interfaces are subsequently defined. Since the initial distribution of the powder particles in the liquid is randomly generated, some particles may share a boundary. If any two particles have different spin index S then the boundary cells are set to be the interface cells. However, if the two particles share the same spin index, they are considered as a single particle. A flow chart of the computational algorithm used for the model is shown in Figure 2 and it can be summarized as follows:

1) Seed sites are randomly chosen and grown until the user specified powder-to-liquid ratio is achieved;

2) Future solute concentrations in the liquid and solid phases are simultaneously calculated for one time step;

3) A cell is randomly chosen, and if it is determined to be liquid or solid then it is left unchanged. However, if it is an interface cell, interface cell velocities in the X and Y directions are calculated;

4) The change in solid fraction, Δf, is calculated to update the existing solid fraction in the interface cell;

5) If the new solid fraction in the interface cell is larger than 1, the solid fraction is set to 1 and the extra solid fraction is evenly distributed among identified neighboring liquid cells. The cells subsequently become interface cells and the interfacial curvature and concentrations is calculated. Likewise, if the solid fraction of the cell is less than 0, the solid fraction is reset to 0. The neighboring solid cells are then determined and the difference in solid fraction is evenly

extracted from the neighboring solid cells. The solute concentration in the neighborhood is also adjusted to account for the change in solid fraction;

6) Cell spin index is updated depending on particle growth or dissolution. A dissolved solid cell has a spin index of zero;

7) Steps 3 - 6 are repeated such that all the cells in the domain are scanned at least once;

8) All data associated with each cell are updated and steps 2 - 7 are repeated for the next time step.

3. RESULTS AND DISCUSSION

To study the influence of bonding temperature on the dissolution behavior of gap-filler powder particles by molten filler alloy, numerical simulations were carried out by using the new model for a Ni-B system at an initial bonding temperature of 1180°C. The liquid phase at the start of the simulation is a molten alloy of nickel and boron with a boron concentration of 17 at%, while the gap-filler powder consists of nickel particles that are initially free of boron. Figure 3 shows the normalized boron distribution in the liquid and the solid particles after 20 time steps for 50 randomly distributed powder particles with an average solid-to-liquid volume ratio of 10%. A number of important observations can be made from the simulations. As can be seen at first glance, after 20 time steps, the liquid is rich in boron while the center region of the particles has the lowest boron concentration while localized depletion of the solute is observed in the liquid at regions closest to the liquid-solid interfaces. Further increase in holding time resulted in further diffusion of boron into the solid particles along with dissolution of the solid particles as shown in Figure 4. Since initial liquid and solid concentrations near the liquidsolid interfaces are different than the local equilibrium concentrations at the bonding temperature, there exists a driving force to reach this equilibrium which results in dissolution of the solid particle to reduce solute concentration in liquid. A larger amount of solute depletion in liquid will exit at regions where density of powder particle distribution is high while regions free of powder particles have high solute concentrations. The decrease in the solute concentration in the liquid to the equilibrium concentration with increase in time, by the dissolution of the gap-filler powder particles, was achieved without the complete dissolution of all gap-filler particles. The incomplete dissolution of the solid particles resulted in the formation stray-grains within the joint region after complete isothermal solidification of the interlayer liquid, as shown in Figure 5. The simulation confirms experimental observations reported in the literature that incomplete dissolution of gap-filler powder particles produces formation of stray grains during TLP bonding of single crystal substrates [3,4].

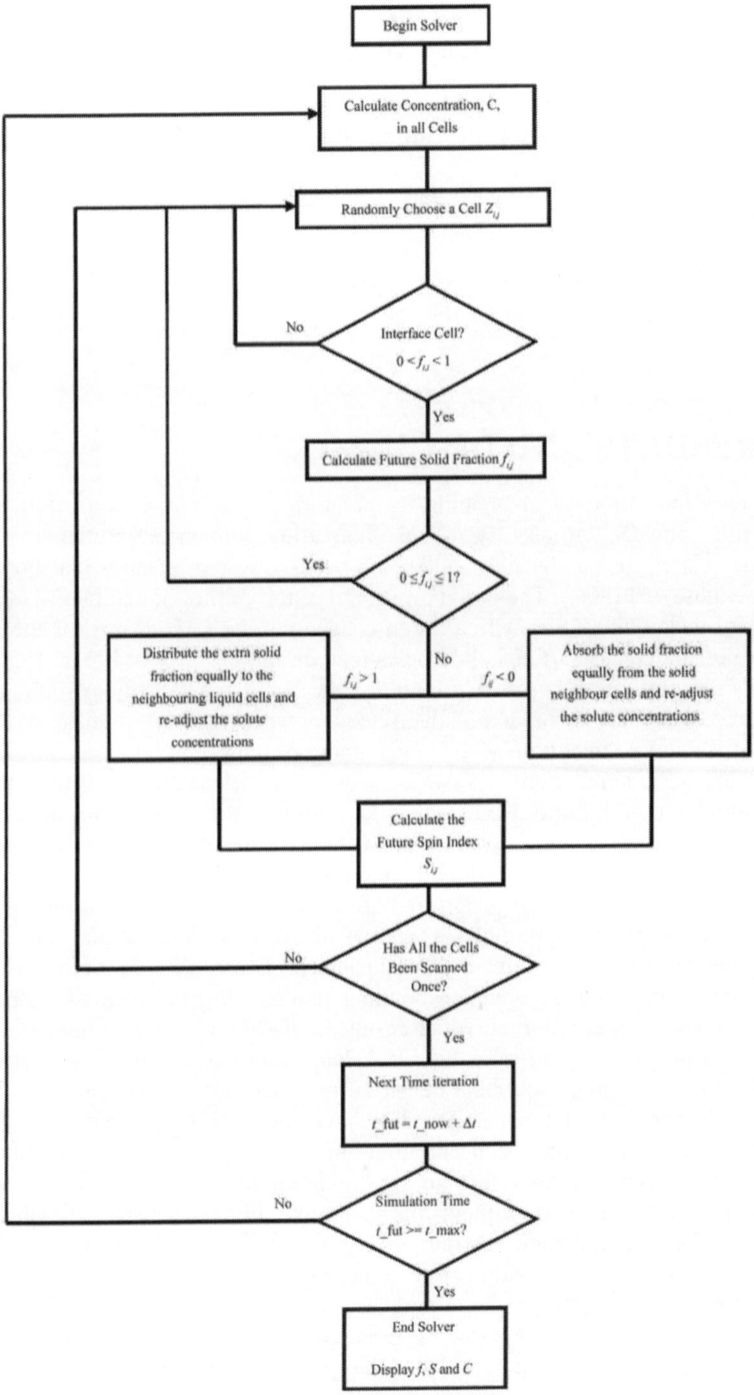

Figure 2. Flow chart of the computational algorithm for the cellular automata model.

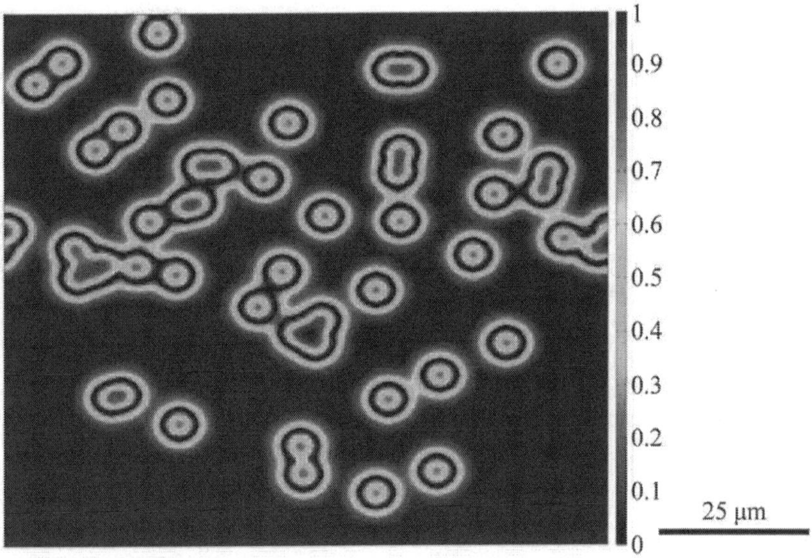

Figure 3. Numerically simulated micrograph of normalized solute distribution in the liquid and gap-filler powder particles after 20 time steps.

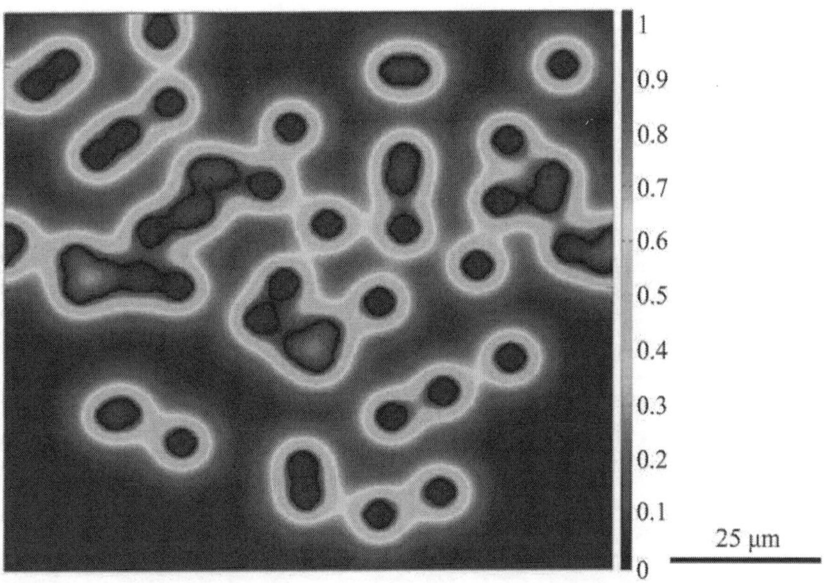

Figure 4. Numerically simulated micrograph of normalized solute distribution in the liquid and gap-filler powder particles after 60 time steps.

The effect of increase in temperature on the gap-filler powder dissolution was simulated by increasing the temperature to 1250°C, while keeping all the other parameters the same. As shown in Figures 6(a)-(c), dissolution of the powder particles gradually occurred until the particles were completely dissolved after 400 time steps. It can also be observed that agglomerated particles will gradually dissociate and breakup from their immediate neighbor during the dissolution process. Detecting this behavior using the conventional interface-tracking methods would have posed many computational difficulties, however, it can be reliably simulated by using the CA method. The complete dissolution of the gap-filler powder particles resulted in a joint with matching crystallographic orientation with the substrate material, without the formation of stray grains (Figure 7). The simulation results, thus, show that in contrast to what is generally reported, complete dissolution of gap-filler powder particles is possible during the bonding process and when it occurs, it produces joint free of stray-grains upon complete isothermal solidification.

Variation of residual powder particles with bonding temperature is illustrated in Figure 8, which shows that an increase in temperature produces increased powder dissolution with complete dissolution occurring at 1250°C. Aside from temperature, other salient bonding parameters that influence the occurrence of complete gap-filler powder dissolution, which is not usually reported, are discussed next.

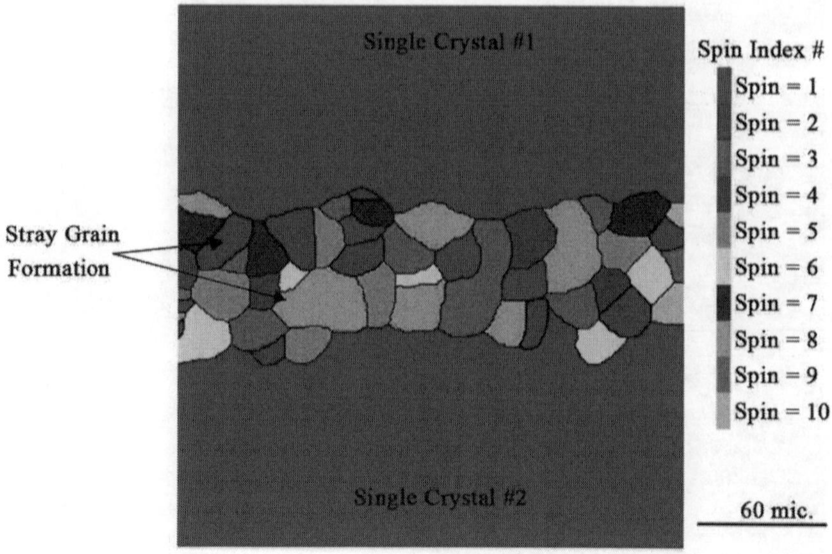

Figure 5. Numerical simulation spin index distribution showing formation of misoriented stray-grains in TLP joint between single crystal substrates.

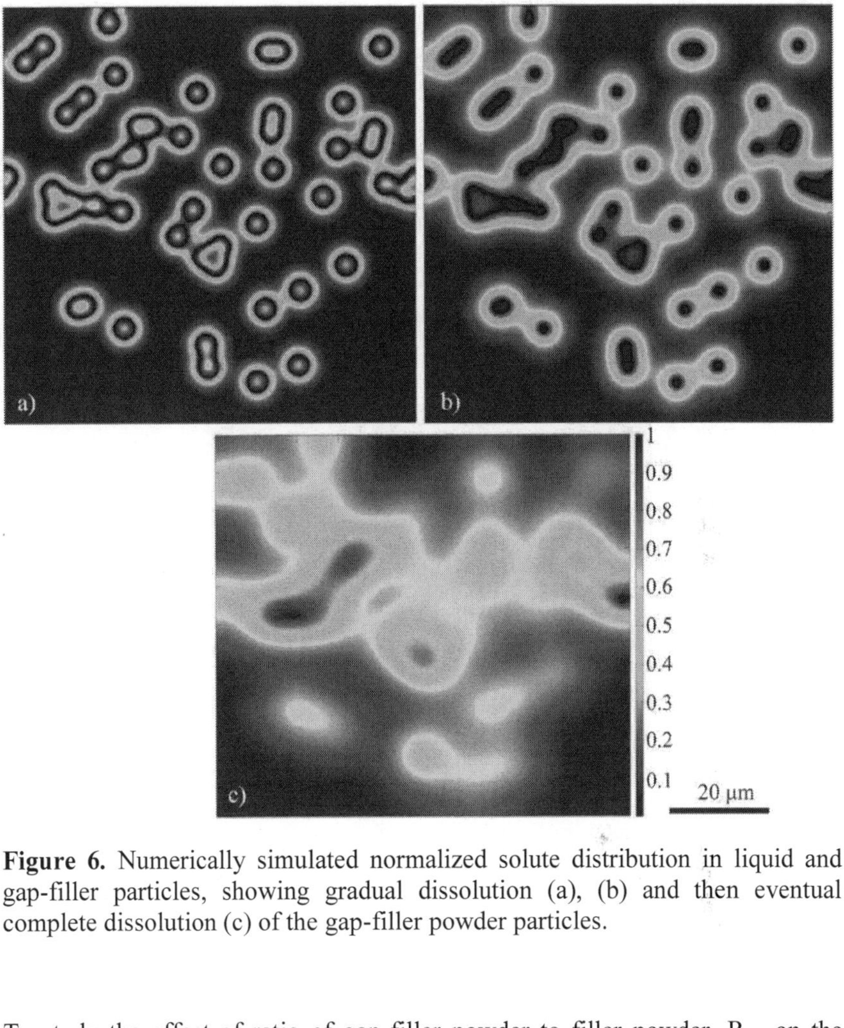

Figure 6. Numerically simulated normalized solute distribution in liquid and gap-filler particles, showing gradual dissolution (a), (b) and then eventual complete dissolution (c) of the gap-filler powder particles.

To study the effect of ratio of gap-filler powder to filler powder, R_{af}, on the extent of dissolution of the gap-filler particles, numerical simulations were performed by varying R_{af} from 10% - 60% at 1250°C bonding temperature. The variations of residual gap-filler particles remaining with time, for the different values of R_{af} are shown in Figure 9. Complete dissolution of powder particles occurred with R_{af} values of 10%, 30% and 40% powder mixtures, while the particles experience incomplete dissolution in the 50% and 60% mixtures. The results show that the R_{af} has a significant influence on the occurrence of partially dissolved powder particles that result in the formation of stray-grains. As the R_{af} increases, the amount of solid present will exceed the amount of solid dissolution required to dilute the molten filler alloy in order to attain the equilibrium solute concentration at the bonding temperature. In addition, as more gap-filler powder is added, the amount of solute rich liquid that requires dilution decreases. These concurrent effects can explain the observation of complete gap-filler powder dissolution with reduction in R_{af}.

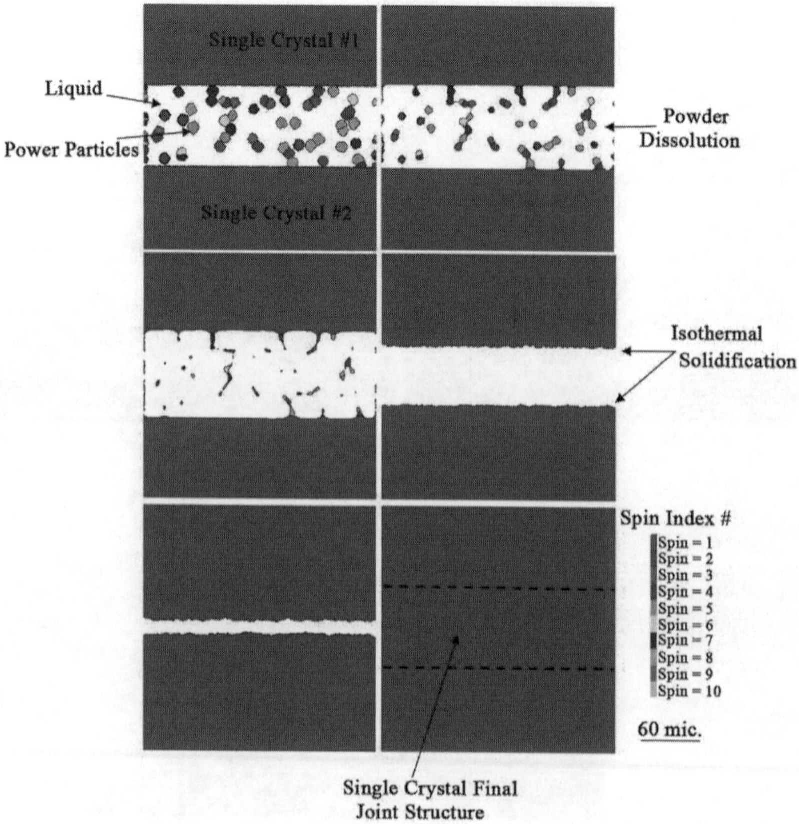

Figure 7. Numerical simulation spin index distribution showing gradual to complete dissolution of the gap-filler powder particles and a final single crystal joint formed after complete isothermal solidification of the interlayer liquid.

Commercial filler alloy powders are produced with varying composition of MPD solute. In the present work, the influence of concentration of MPD solute in the filler alloy powder, C_F, on dissolution of gap-filler powder particles was also investigated. The simulation was performed with a R_{af} value of 30% and at a bonding temperature of 1100°C. As shown in Figure 10, the extent of the gap-filler powder dissolution increases with increase in C_F. At a given temperature, an increase in the initial concentration of the MPD solute in the filler alloy raises the amount of dilution of the molten filler required for the liquid to attain the equilibrium composition, which in turn increases the dissolution of the gap-filler powder particles.

Hence, the numerical calculations performed in this work indicate that, even though it is not generally reported, depending on the combination of various factors, complete dissolution of the gap-filler powder particles by molten filler alloy can occur during TLP bonding. The CA model also shows that under such a situation of complete liquid phase dissolution of the gap-filler powder particles, the use of composite powder mixture can be used to produce stray-grains-free joint with matching crystallographic orientation with the substrate material. The model predictions were experimentally verified by using a commercial Ni-Cr-B filler alloy powder (Nicrobraz 150) and IN 738 superalloy as gap-filler powder with IN 738 superalloy as the substrate material. The chemical compositions of the filler alloy and IN 738 are listed in Table 1. The experiments were performed at different temperatures using a composite powder mixture of the filler alloy and gap-filler alloy with a R_{af} value of 30%. At temperatures below 1150°C, the molten filler alloy only caused partial melting of the gap-filler powder particles. A micrograph of an incompletely melted power mixture is shown in Figure 11. An increase in the temperature to 1150°C resulted in complete liquid-phase dissolution of the gap-filler particles by the molten filler alloy (Figure 12), Complete isothermal solidification of the fully melted composite powder mixture produced a joint that is free of stray-grains between single crystal IN 738 substrates (Figure 13).

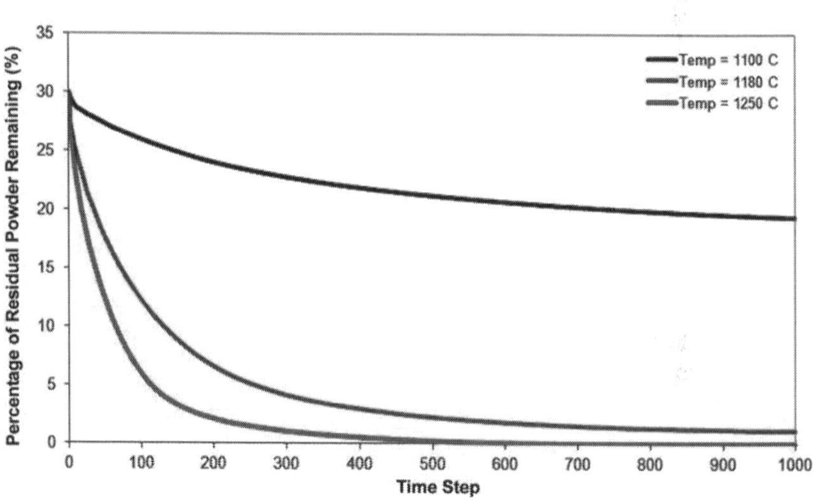

Figure 8. Numerical simulation result of gap-filler particles dissolution behavior at different holding temperatures.

Orientation imaging microscopy (OIM) analyses were performed on the complete isothermally solidified joint produced with the composite powder, to evaluate the shows {100}, {110} and {111} stereographic pole figures of randomly chosen nine points within the substrate crystallographic orientation relationship between the joint and the single crystal substrate and to confirm the preclusion of stray-grains within the joint. Figure 14 and joint region. The analysis shows that the nine points are projected at nearly the same location within the stereographic pole figures. Also, the result of an OIM mapping technique that uses color to represent crystallographic orientation is presented in Figure 15. The results of both OIM techniques (Figures 14 and 15) confirm that the SX substrate material and the joint region exhibit matching crystallographic orientations. The experimental results, thus, corroborate the numerical model prediction that indicates complete melting of powder mixture of filler alloy and gap-filler alloy is possible during TLP bonding and complete isothermal solidification of the liquid produced by such fully melted mixture can prevent forma- tion of stray grains during joining of single crystals.

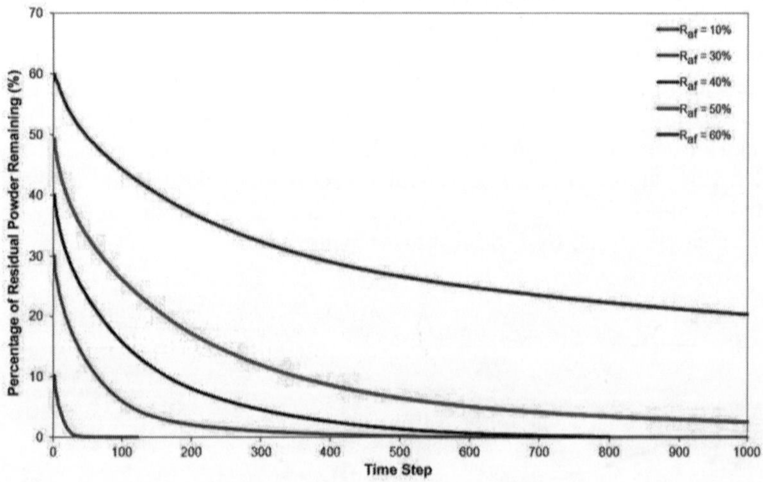

Figure 9. Numerical simulation result showing the effect of gap-filler powder to filler powder ratio on the gap-filler particles dissolution behavior.

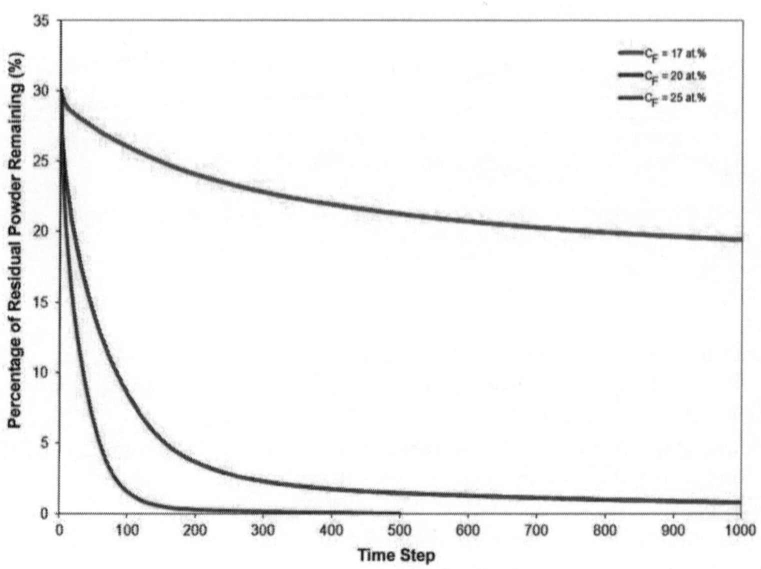

Figure 10. Numerical simulation result showing the effect of initial MPD solute concentration in liquid on gap-filler powder dissolution behavior.

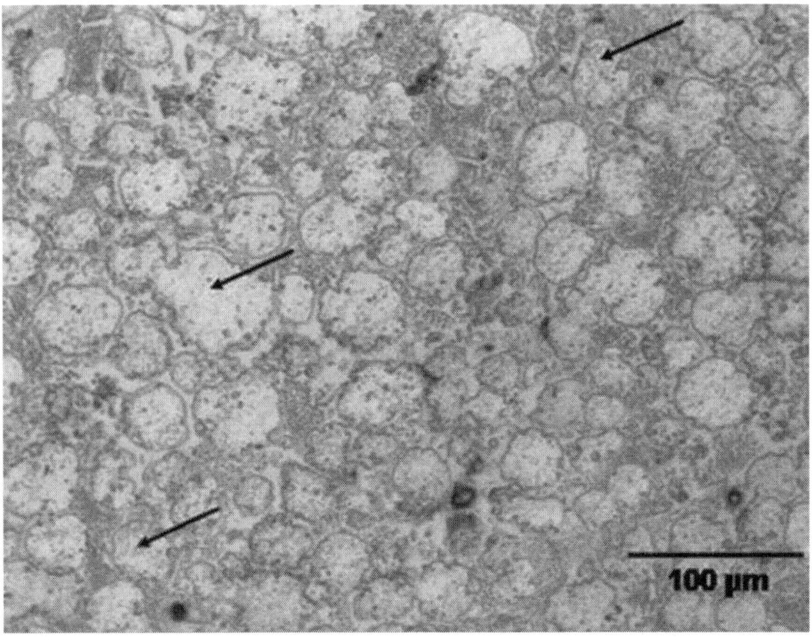

Figure 11. Microstructure of a partially melted powder mixture. The arrows show residual gap-filler powder particles.

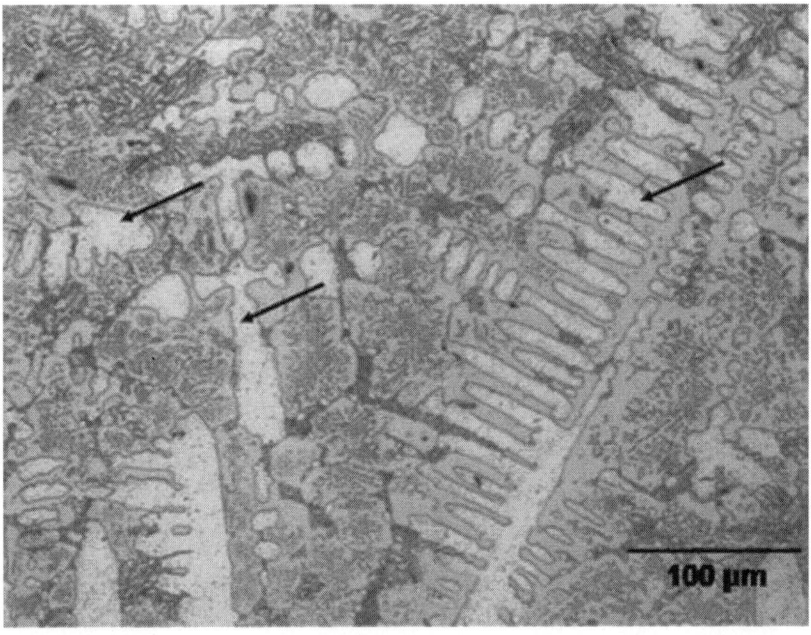

Figure 12. Microstructure of completely melted interlayer powder mixture. The arrows show dendrites formed after solidification.

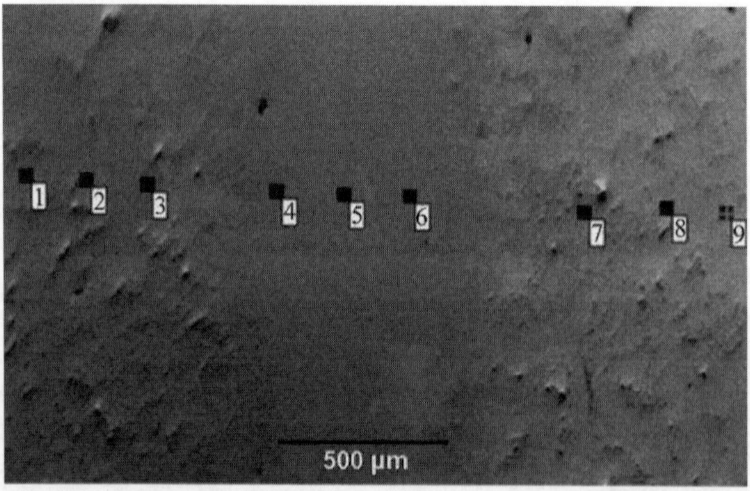

Figure 13. Micrograph of a joint free of stray-grains between single crystal IN 738 substrates produced with composite powder mixture with a R_{af} value of 30% at 1150°C.

4. SUMMARY AND CONCLUSIONS

1) A new CA numerical model has been developed to study the dissolution of solid gap-filler powder particles by molten filler alloy during the TLP bonding process.

2) The numerical analysis shows that the liquid-phase dissolution of the powder particles is significantly influenced by temperature, the initial concentration of MPD solute in the filler alloy and the ratio of the gap-filler alloy to the filler alloy in the composite powder mixture.

3) The use of spin index, S, in the theoretical model, to represent the crystallographic orientation of each powder particle shows that incomplete dissolution of the gapfiller particles would result in the formation of misoriented stray-grains in a joint between single crystal substrates.

4) In contrast to what has been generally assumed and reported, numerical calculations show that complete dissolution of the gap-filler particles by the molten filler is possible and complete isothermal solidification of the resultant liquid would produce a joint having a matching crystallographic orientation with single crystal substrate.

5) The predictions of the numerical model were experimentally verified and complete melting of a composite powder mixture was observed and it produced a joint free of misoriented stray-grains in a single crystal IN 738 superalloy substrate.

Table 1. Nominal materials composition.

Material	Ni	Cr	Co	Mo	W	Nb	Al	Ti	Fe	Ta	C	B	Zr	S
IN738	Bal.	15.84	8.5	1.88	2.48	0.92	3.46	3.47	0.07	1.69	0.11	0.012	0.04	0.001
Nicrobraz 150	Bal	15	---	---	---	---	---	---	---	---	0.03	3.5	---	---

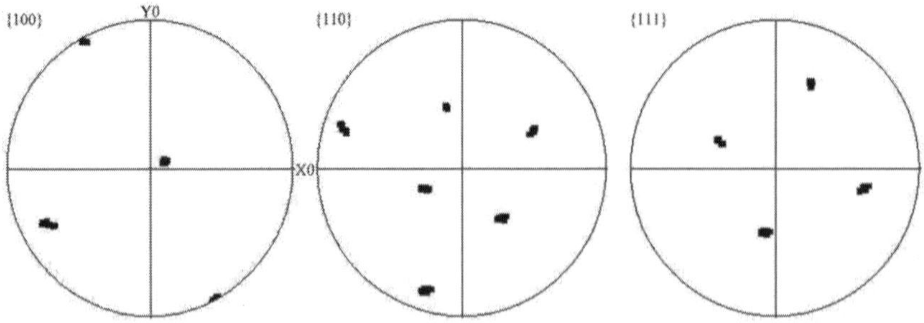

Figure 14. {100}, {110} and {111} stereographic pole figures of the analyzed locations shown in Figure 12.

Figure 15. Result of orientation imaging microscopy mapping across the SX IN 738 substrates and the joint region shown in Figure 12. The same color of the joint region and the substrates indicates matching crystallographic orientation relationship.

ACKNOWLEDGEMENTS

The authors gratefully acknowledge financial support by NSERC of Canada.

REFERENCES

1. R. A. MacKay, R. L. Dreshfield and R. D. Maier, "Anisotropy of Nickel-Base Superalloy Single Crystals," Proceedings of the 4th International Symposium for Superalloys, Seven Springs, Champion, 21-25 September 1980, pp. 385-394.

2. D. S. Duvall and W. A. Owczarski, "Further Heat-AffectedZone Studies in Heat Resistant Nickel Alloys," Welding Journal, Vol. 46, No. 9, 1967, pp. 423-432.

3. W. F. Gale and D. A. Butts, "Transient Liquid Phase Bonding," Science and Technology of Welding Journal, Vol. 9, No. 4, 2004, pp. 283-300.

4. P. C. Prakash, A. Merati, X. Huang and J.-F. Thibault, "Advanced Technologies for Repair of Single Crystal Turbine Blades," Proceeding of 43rd Conference of 450 Metallurgist, Hamilton, 22-25 August 2004, p. 413.

5. D. S. Duvall, W. A. Owczarski and D. F. Paulonis, "Transient Liquid Phase Bonding: A New Method for Joining Heat Resistant Alloys," Welding Journal, Vol. 53, No. 4, 1974, pp. 203-214.

6. P. S. Kotval, J. D. Venables and R. W. Calder, "The Role of Hafnium in Modifying the Microstructure of Cast Nickel-Base Superalloys," Metallurgical and Materials Transactions B, Vol. 3, No. 2, 1972, pp. 457-462.

7. J.-H. Huang, Y.-L. Dong, Y. Wan, X.-K. Zhao and H. Zhang, "Investigation on Reactive Diffusion Bonding of SiCp/6063 MMC by Using Mixed Powders as Interlayers," Journal of Materials Processing Technology, Vol. 190, No. 1-3, 2007, pp. 312-316.

8. X. W. Wu, R. S. Chandel, H. P. Seow and H. Li, "Wide Gap Brazing of Stainless Steel to Nickel-Based Superalloy," Journal of Materials Processing Technology, Vol. 113, No. 1-3, 2001, pp. 215-221.

9. W. D. Zhuang and T. W. Eager, "Liquid Infiltrated Powder Interlayer Bonding: A Process for Large Gap Joining," Science and Technology of Welding and Joining, Vol. 5, No. 3, 2000, pp. 125-135.

10. Y. Zhou, "Analytical Modeling of Isothermal Solidification during Transient Liquid Phase (TLP) Bonding," Journal of Materials Science Letters, Vol. 20, No. 9, 2001, pp. 841-844.

11. Y. Zhou, W. F. Gale and T. H. North, "Modeling of Transient Liquid Phase Bonding," International Materials Review, Vol. 40, No. 5, 1995, pp. 181-196.

12. Y. Nakao, K. Nishimoto, K. Shinozaki and Y. Kang, "Transient Liquid Insert Metal Diffusion Bonding of NickelBase Superalloys," Joining Advanced Materials, Chapman and Hall, London, 1990, pp. 129-144.

13. Tuah-Poku, M. Dollar and T. B. Massalski, "A study of the Transient Liquid Phase Bonding Process Applied to an Ag/Cu/Ag Sandwich Joint," Metallurgical and Materials Transactions A, Vol. 19, No. 3, 1988, pp. 675-686.

14. Y. Zhou and T. H. North, "Kinetic Modeling of Diffusion-Controlled, Two-Phase Moving Interface Problems," Modeling and Simulation in Materials Science and Engineering, Vol. 1, No. 4, 1993, pp. 505-516.

15. T. C. Illingworth and I. O. Golosnoy, "Numerical Solutions of Diffusion-Controlled Moving Boundary Problems Which Conserve Solute," Journal of Computational Physics, Vol. 209, No. 1, 2005, pp. 207-225.

16. J. F. Li, P. A. Agyakwa and C. M. Johnson, "A fixedGrid Numerical Modeling of Transient Liquid Phase Bonding and Other Diffusion-Controlled Phase Changes," Journal of Materials Science, Vol. 45, No. 9, 2010, pp. 2340-2350.

17. S. Ulam, "Random Processes and Transformations," Proceedings of International Congress of Mathematicians, Cambridge, 30 August-6 September 1950, pp. 264-275.

18. S. Wolfram, "Universality and Complexity in Cellular Automata," Physica D: Nonlinear Phenomena, Vol. 10, No. 1-2, 1984, pp. 1-35.

19. K. Janssens, "An Introductory Review of Cellular Automata Modeling of Moving Grain Boundaries in Polycrystalline Materials," Mathematics and Computers in Simulation, Vol. 80, No. 7, 2010, pp. 1361-1381.

20. H. Yang, C. Wu, H. W. Li and X. G. Fan, "Review on Cellular Automata Simulations of Microstructure Evolution during Metal Forming Process: Grain coarsening, recrystallization and phase transformation," Science China Technological Sciences, Vol. 54, No. 8, 2011, pp. 2107- 2118.

21. Y. Liu, Q. Y. Xu and B. C. Liu, "A Modified Cellular Automaton Method for the Modeling of the Dendritic Morphology of Binary Alloys," Tsinghua Science and Technology, Vol. 11, No. 5, 2006, pp. 495-500.

22. W. Wang, P. D. Lee and M. McLean, "A Model of Solidificaiton Microstructure in Nickel-Based Superalloys: Predicting Primary Dendrite Spacing Selection," Acta Materialia, Vol. 51, No. 10, 2003, pp. 2971-2987.

23. K. Kremeyer, "Cellular Automata Investigation of Binary Solidification," Journal of Computational Physics, Vol. 142, No. 1, 1998, pp. 243-262.

24. S. Raghavan and S. Sahay, "Modeling the Grain Growth Kinetics by Cellular Automaton," Materials Science and Engineering A, Vol. 445, 2007, pp. 203-209.

25. X. L. Yang, H. B. Dong, W. Wang and P. D. Lee, "Microscale Simulation of Stray Grain Formation in Investment Cast Turbine Blades," Materials Science and Engineering A, Vol. 386, No. 1-2, 2004, pp. 129-139.

26. M. J. Krane, D. R. Johnson and S. Raghavam, "The Development of Cellular Automaton-Finite Volume Model for Dendritic Growth," Applied Mathematical Modelling, Vol. 33, No. 5, 2009, pp. 2234-2247.

27. C.-A. Gandin, J.-L. Desbiolles, M. Rappaz and Ph. Thevoz, "A Three-Dimensional Cellular Automaton-Finite Element Model for the Prediction of Solidification Grain Structures," Metallurgical and Materials Transactions A, Vol. 30A, No. 12, 1999, pp. 3153-3165.

28. Ch.-A. Gandin and M. Rappaz, "A Coupled Finite Element-Cellular Automaton Model for the Prediction of Dendritic Grain Structure in Solidification Processes," Acta Metallurgica et Materialia, Vol. 42, No. 7, 1994, pp. 2233-2246.

29. H. Yin, S. D. Felicelli and L. Wang, "Simulation of a Dendritic Microstructure with the Lattice Boltzmann and Cellular Automaton Methods," Acta Materialia, Vol. 59, No. 8, 2011, pp. 3124-3136.

30. R. Sasikumar and R. Sreenivasan, "Two-Dimensional Simulation of Dendrite Morphology," Acta Metallurgica et Materialia, Vol. 42, No. 7, 1994, pp. 2381-2386.

31. D. M. Li, R. Li and P. W. Zhang, "A Cellular Automaton Technique for the Modeling of Solidification Microstructure in Multi-Component Alloys," 1st International MultiSymposiums on Computer and Computational Sciences, Hangzhou, 20-24 June 2006, pp. 800-806.

32. Ghoneim and O. A. Ojo, "Numerical Modeling and Simulation of Diffusion-Controlled Liquid-Solid Phase Change in Polycrystalline Solids," Computational Materials Science, Vol. 50, No. 3, 2011, pp. 1102-1113.

CHAPTER 9

Synthesis of 1,4-Bis(phenylethynyl)benzenes and Their Application as Blue Phase Liquid Crystal Composition

Ning Li [2,3], Zhengqiang Li [2,3], Xing Zhang [2,3] and Ruimao Hua [1,2,3,*]

[1] Department of Chemistry, Tsinghua University, Beijing 100084, China

[2] Beijing R&D Center, Shijiazhuang Chengzhi Yonghua Display Materials Co. Ltd., Beijing 100083, China

[3] Hebei Engineering & Technology Center for FPD Materials, Shijiazhuang 050091, China

ABSTRACT

A number of 1,4-bis(phenylethynyl)benzene derivatives (BPEBs) and their analogues with different numbers of side-substitute fluorine atoms on benzene rings, and alkyl chains, ethoxyl groups, fluorine atoms and trifluoromethyl groups as the end groups have been synthesized. The effects of the different substituents on their properties such as thermal behavior of melting point and clearing point, the temperature of nematic phase, optical anisotropy and dielectric anisotropy have been well investigated, and it has been found that some BPEBs have a wide range of the nematic phase temperature with high optical anisotropy (Δn) and acceptable dielectric anisotropy ($\Delta\varepsilon$), which have been applied as the crucial compositions to constitute a liquid crystal mixture having the properties of $\Delta\varepsilon = 29.0$ and $\Delta n = 0.283$ at 25 °C. With the addition of the chiral dopant to the obtained liquid crystal mixture, blue phase liquid crystal with a blue phase temperature range of 8 °C has been achieved.

KEYWORDS

1,4-bis(phenylethynyl)benzene derivatives; blue phase; liquid crystal

1. INTRODUCTION

Liquid crystals (LCs) have had a multitude of applications in the past few decades [1,2]. One of the important and unique applications is their use as the key fundamental materials to develop LC displays [3], which have actually changed people's lifestyle due to the use of mobiles, notebook computers, flat panel desktop monitors, and LCD televisions, etc. The development of excellent LC displays, which have the advantages of fast response, high contrast ratio, and low driving voltage, depends greatly on the development of new types and properties of LCs. The demand for LCDs with a fast response is one of the crucial factors to improve the quality of the displays, and blue phase liquid crystal (BPLC) is commonly considered to be one of the strongest candidates [4–7].

On the other hand, the molecular and electronic structures of 1,4-bis(phenylethynyl)-benzene derivatives (BPEBs), as well as their applications have attracted much attention recently [8–12]. In particular, BPEBs have been applied as the important components in LCs with the characters of high melting point (mp), clearing point (cp), and large optical anisotropy values (Δn) [13–22]. It is important and interesting to investigate the substituent effect on the properties of BPEBs. Therefore, in this paper, we describe the synthesis of a number of BPEBs with different alkyl chains, the numbers of fluorine atoms and other substituents in benzene rings as shown in Scheme 1, and the study on the substituent effects on their properties, as well as their application as blue phase liquid crystal composition.

2. RESULTS AND DISCUSSION

2.1. Synthesis of BPEBs and Analogues (1–25)

The synthetic routes of **1–25** are outlined in Scheme 2, including the key steps of the formation of terminal and internal alkynes via Sonogashira cross-coupling reactions of aryl iodides/bromides catalyzed by palladium(0) complexes in good to high yields, and a typical synthetic procedure for the formation of **12** is described in the Experimental Section (vide infra).

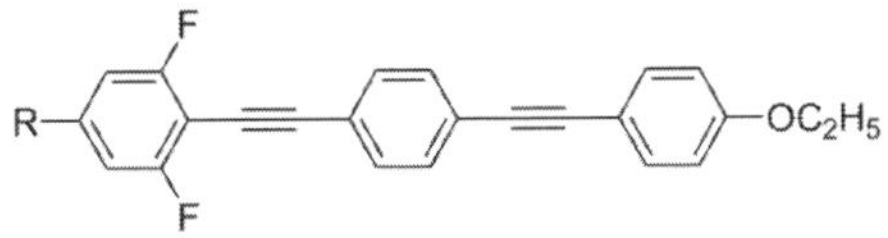

$R = C_2H_5$ (**1**); $R = n\text{-}C_3H_7$ (**2**);
$R = n\text{-}C_4H_9$ (**3**); $R = n\text{-}C_5H_{11}$ (**4**)

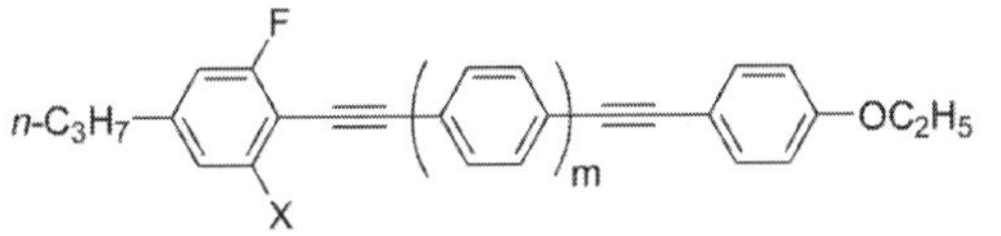

X = H, m = 1 (**5**)
X = H, m = 2 (**6**)
X = F, m = 2 (**7**)

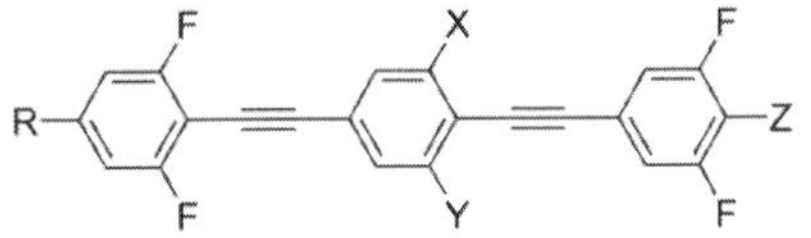

$R = n\text{-}C_3H_7$, X = Y = H, Z = OCF_3 (**8**)
$R = n\text{-}C_3H_7$, X = Y = H, Z = F (**9**)
$R = n\text{-}C_3H_7$, X = F, Y = H, Z = OCF_3 (**10**)
$R = n\text{-}C_3H_7$, X = Z = F, Y = H (**11**)
$R = n\text{-}C_3H_7$, X = Y = F, Z = OCF_3 (**12**)
$R = n\text{-}C_3H_7$, X = Y = Z = F (**13**)

$R = n\text{-}C_4H_9$, X = Y = H, Z = OCF_3 (**14**)
$R = n\text{-}C_4H_9$, X = Y = H, Z = F (**15**)
$R = n\text{-}C_4H_9$, X = F, Y = H, Z = OCF_3 (**16**)
$R = n\text{-}C_4H_9$, X = Z = F, Y = H (**17**)
$R = n\text{-}C_4H_9$, X = Y = F, Z = OCF_3 (**18**)
$R = n\text{-}C_4H_9$, X = Y = Z = F (**19**)

$R = n\text{-}C_5H_{11}$, X = Y = H, Z = OCF_3 (**20**)
$R = n\text{-}C_5H_{11}$, X = Y = H, Z = F (**21**)
$R = n\text{-}C_5H_{11}$, X = F, Y = H, Z = OCF_3 (**22**)
$R = n\text{-}C_5H_{11}$, X = Z = F, Y = H (**23**)
$R = n\text{-}C_5H_{11}$, X = Y = F, Z = OCF_3 (**24**)
$R = n\text{-}C_5H_{11}$, X = Y = Z = F (**25**)

Scheme 1. The structures of 1,4-bis(phenylethynyl)benzene derivatives and analogues (**1**–**25**).

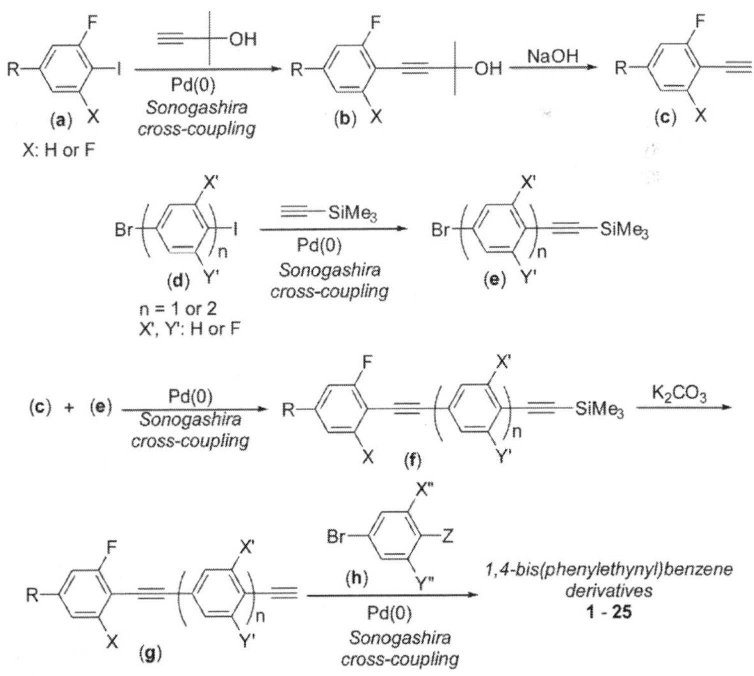

Scheme 2. Synthesis of 1,4-bis(phenylethynyl)benzene derivatives and analogues.

2.2. Thermal Properties

The thermal properties of the melting point (mp) and clearing point (cp) are critical in the practical utilization of the synthesized BPEBs as the composition of LCs, thus the mp and cp were determined by DSC (Differential Scanning Calorimetry), and their thermal data as well as enthalpic data (ΔH) are concluded in Table 1. It was found that mp and cp greatly depended upon the molecular and electronic structures of BPEBs. BPEBs **1–4** clearly show the effect of alkyl chain length on mp and cp, and when n-C_3H_7 and C_2H_5 groups are used as the end groups, BPEBs **1** and **2** give the similar mp and cp (**1**vs.**2**). However, when the longer alkyl chains of C4 and C5 were employed, BPEBs **3** and **4** showed similar thermal properties, but both mp and cp decreased greatly (**1** & **2**vs.**3** & **4**) [23]. It can be concluded that BPEBs **3** and **4** bearing the longer alkyl chains have a wider range of nematic phase temperature than those of BPEBs **1** and **2**, possibly due to the longer alkyl chains being more flexible than short alkyl chains. Both **3** and **4** show a nematic phase temperature range of about 140 °C, and the wide nematic phase temperature are very important to make a practical LC mixture.

By comparison of **2** and **5**, it was found that decreasing the number of fluorine atoms in the end benzene ring leads to the increase of both mp and cp in a range of about 10 °C, and a similar trend of mp was also found between 4,4′-bis(phenylethynyl)biphenyls **6** and **7**, which are the analogues of BPEBs **2** or **5**.

The thermal properties of other BPEBs are also compared to each other, as shown in Figure 1, and two notable features of the relationship between mps and their chemical structure could be concluded as follows: (1) The introduction of the fluorine atom on the middle benzene ring shows a great effect on the change of mp. In general, one or two fluorine atom-substituted benzene result in the increase of mp, and the introduction of the second fluorine atom leads to much more significant increase of mp relative to the first fluorine atom introduction (e.g., **8→10→12; 9→11→13**). (2) When the end group of OCF_3 is replaced by a fluorine atom, the mps decreased (e.g., **10**vs.**11; 12**vs.**13; 14**vs.**15**), and only one exception (**8**vs.**9**) was observed.

2.3. Nematic Phase

Among the synthesis of BPEBs and analogues, only BPEBs **1–5, 8, 20–21** (Figure 2, on heating run) and **9, 14–15** (on cooling run) show nematic phases under polarizing microscope with the structural character without side-substituted fluorine atoms bonded to the middle benzene ring, and the analogues of BPEBs **6** and **7** have no nematic phase either. It was found that the nematic phase temperature is quite different depending on the end groups and the numbers of fluorine atoms. As shown in Figure 3 and Table 1, BPEBs **1–5** with OC_2H_5 as the end group possess wide nematic phase temperatures from 121.6 to 147.0 °C, and have apparently disclosed that longer alkyl chains generally result in a wider temperature of nematic phase. BPEBs **3** and **4** bearing the longer alkyl chain of n-C_4H_9 or n-C_5H_{11} give the maximum nematic phase temperatures, while BPEBs **8**, **20** and **21** with OCF_3 or F as the end groups show the relatively narrow nematic phase temperature range.

Table 1. Melting points and clearing points of 1,4-bis(phenylethynyl)benzene derivatives (BPEBs) and analogues by DSC.

BPEB	Structure	*Mp* (°C)	Δ*H* (J/g)	*Cp* (°C)	Δ*H* (J/g)
1	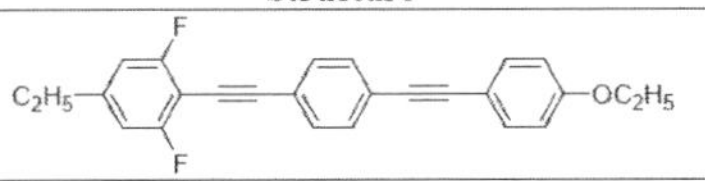	129.5	62.9	251.2	3.0
2	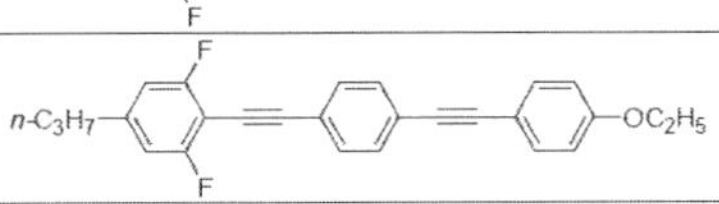	130.2	67.8	254.2	3.8
3	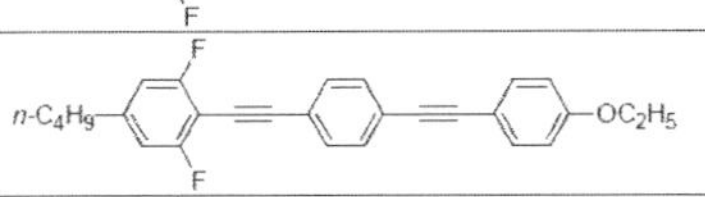	91.0	63.7	238.0	3.8
4	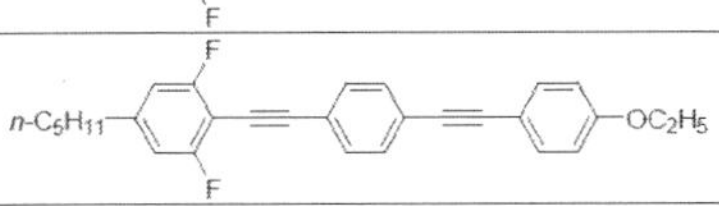	88.7	55.7	232.4	3.8
5	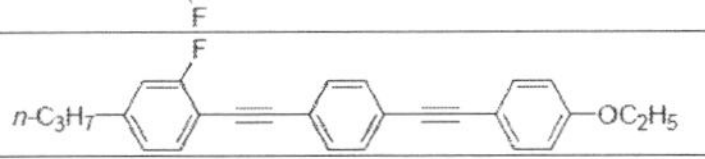	140.4	46.0	262.0	3.0
6		213.2	41.2	–	–
7	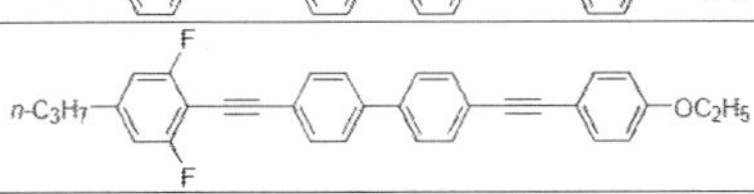	190.4	57.3	–	–
8	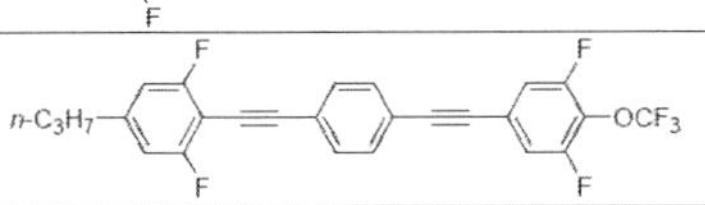	131.1	59.8	152.9	5.8
9	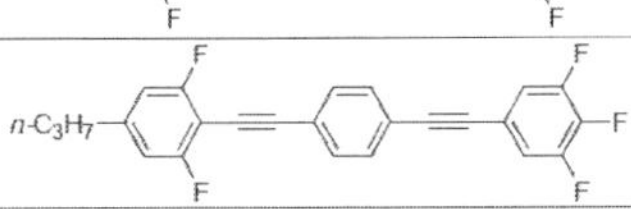	141.4	88.5	[a]	–
10	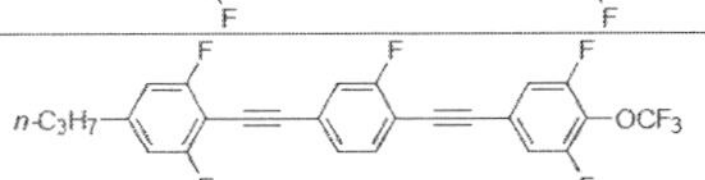	150.7	68.7	–	–
11	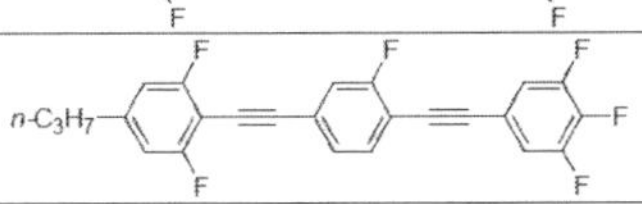	142.8	77.4	–	–
12	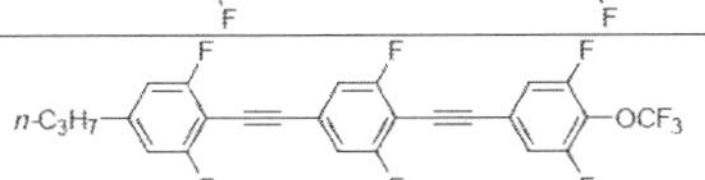	178.2	54.0	–	–

Table 1. *Cont.*

BPEB	Structure	*Mp* (°C)	ΔH (J/g)	*Cp* (°C)	ΔH (J/g)
13	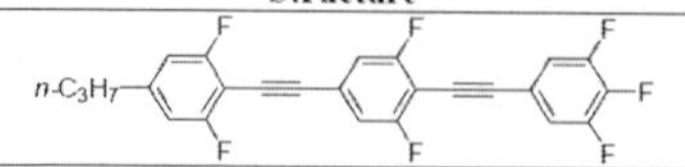	176.4	66.5	–	–
14	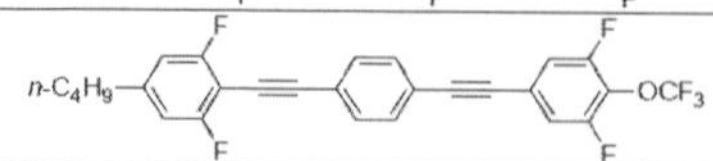	142.6	69.3	b	–
15	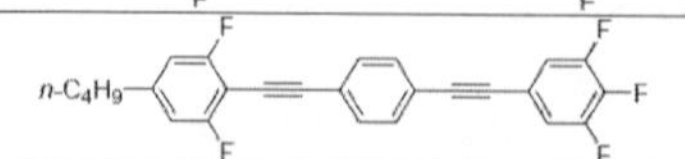	125.2	73.6	c	–
16	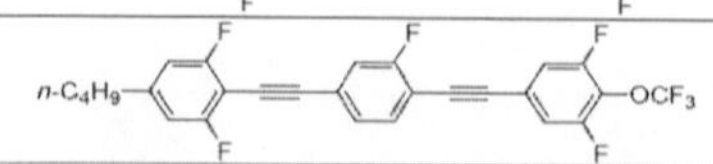	147.8	45.7	–	–
17	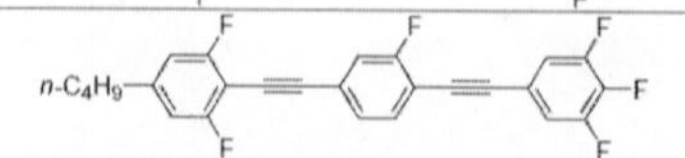	136.2	55.4	–	–
18	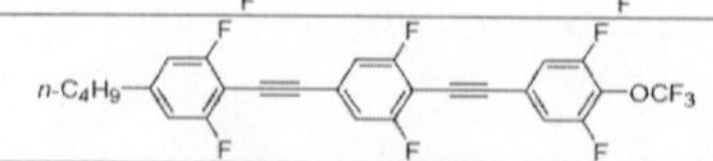	178.3	64.6	–	–
19	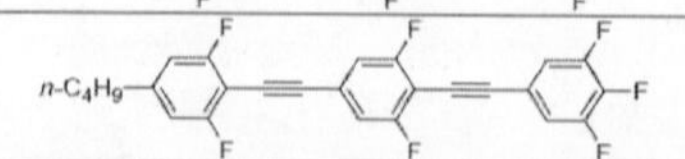	170.8	55.3	–	–
20	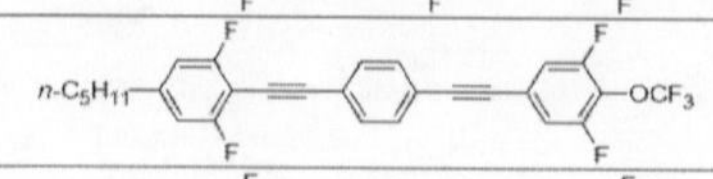	131.3	59.7	144.3	0.81
21	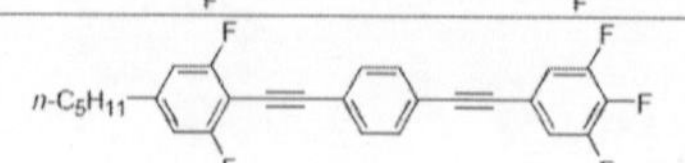	107.7	62.5	122.3	0.29
22	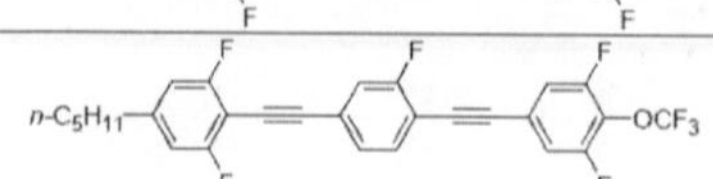	131.3	41.2	–	–
23	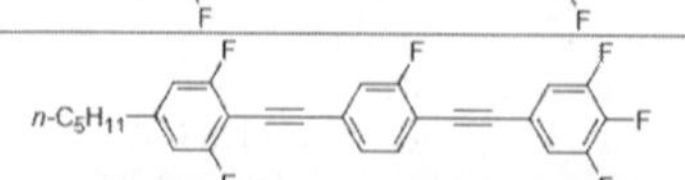	126.0	51.7	–	–
24	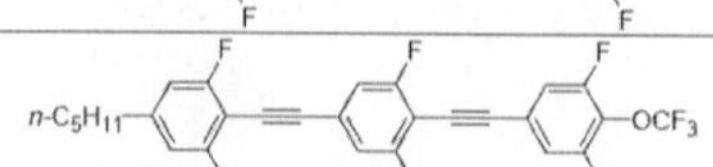	171.1	66.8	–	–
25	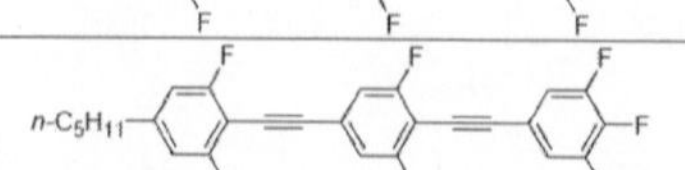	158.7	66.0	–	–

[a] The nematic phase was only observed during cooling in the temperature range 118.3 to 89.3 °C, and iso-*N* 118.3 °C (ΔH = 0.25 J/g), N-Cr 89.3 °C (ΔH = 60.4 J/g); [b] The nematic phase was only observed during cooling in the temperature range 139.6 to 121.4 °C, and iso-*N* 139.6 °C (ΔH = 0.71 J/g), N-Sm 121.4 °C (ΔH = 0.61 J/g), Sm-Cr 113.8 °C (ΔH = 60.27 J/g); [c] The nematic phase was only observed during cooling in the temperature range 107.0 to 71.4 °C, and iso-*N* 107.0 °C (ΔH = 0.16 J/g), N-Cr 71.4 °C (ΔH = 59.1 J/g).

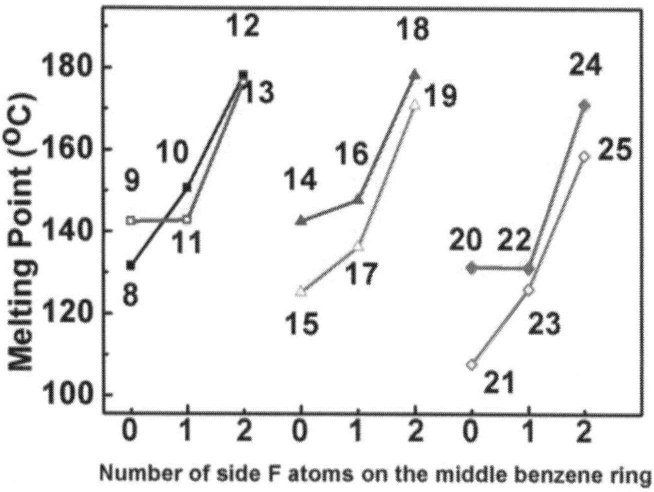

Figure 1. Effect of side-substituted fluorine atoms and end groups on melting points.

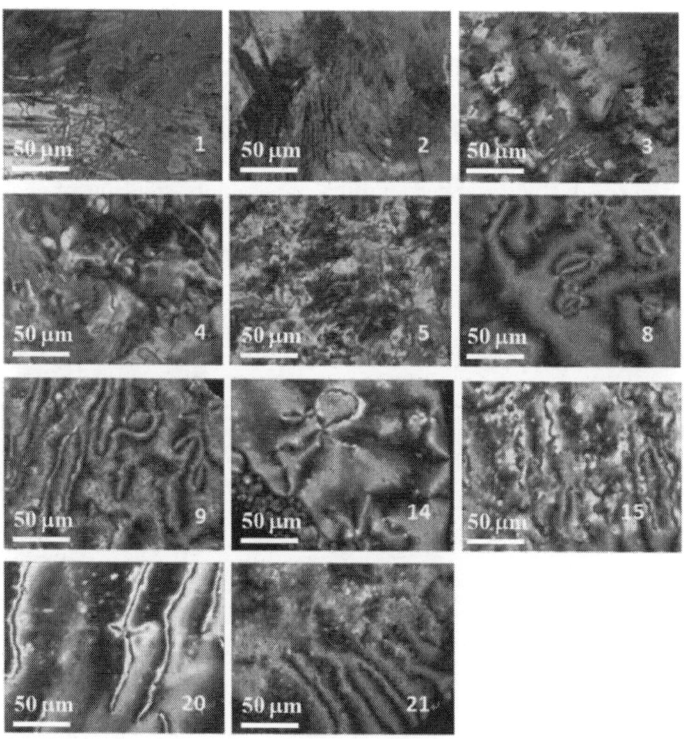

Figure 2. Polarizing optical micrographs photos of BPEBs **1–5**, **8**, **9**, **14**, **15**, **20** and **21**.

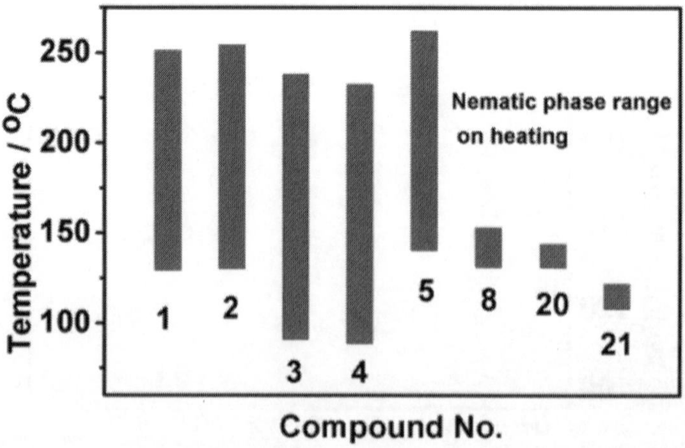

Figure 3. Nematic phase temperature range.

2.4. Optical Anisotropy (Δn)

The optical anisotropy or birefringence (Δn) of the BPEBs showing nematic phase in host LC was determined, and the obtained results were concluded in Table 2. As expected, a large π-conjugated structure leads to relatively high Δn value, and therefore BPEBs **1–5** have higher Δn values than **8–9**, **14–15**, and **20–21**. It is reasonable to understand that much more side-substituted fluorine atoms in different benzene rings will decrease the integrity of π-conjugated structures to result in a decrease of optical anisotropy.

Table 2. Δn values of BPEBs.

BPEB	n_e	n_o	Δn	BPEB	n_e	n_o	Δn
1	2.027	1.531	0.496	**9**	1.867	1.521	0.346
2	1.987	1.511	0.476	**14**	1.797	1.468	0.329
3	1.947	1.511	0.436	**15**	1.867	1.521	0.346
4	2.007	1.511	0.496	**20**	1.807	1.491	0.316
5	2.103	1.521	0.582	**21**	1.867	1.511	0.356
8	1.837	1.491	0.346				

n_e: Extraordinary refraction index; n_o: Ordinary refraction index.

2.5. Dielectric Anisotropy (Δε)

Table 3 shows the values of dielectric anisotropy (Δε) of the BPEBs having nematic phase in host LC. It was found that with the number increase of side-substituted fluorine atoms, BPEBs **8–9**, **14–15**, and **20–21** have higher dielectric anisotropy values than BPEBs **1–5**. In addition, comparison of BPEBs with the end groups of OCF_3 (**8**, **14**, **20**) and F (**9**, **15**, **21**), BPEBs **8**, **14** and **20** have relatively higher dielectric anisotropy values, possibly due to the stronger electronegativity of the OCF_3 group relative to the fluorine atom.

Table 3. Δε values of BPEBs.

BPEB	$\varepsilon_{//}$	$\varepsilon_{\perp}$	Δε	BPEB	$\varepsilon_{//}$	$\varepsilon_{\perp}$	Δε
1	8.9	3.4	5.5	**9**	26.1	5.3	20.8
2	8.0	3.0	5.0	**14**	29.5	6.0	23.5
3	7.2	3.0	4.2	**15**	26.4	5.4	21.0
4	8.1	3.0	5.1	**20**	26.1	5.2	20.9
5	7.3	3.2	4.1	**21**	25.0	5.0	20.0
8	29.5	5.5	24.0				

$\varepsilon_{//}$: Parallel dielectric constant; $\varepsilon_{\perp}$: Vertical dielectric constant.

2.6. Applications of BPEBs as Blue Phase Liquid Crystal Composition

Because the synthesized some of BPEBs have high Δn and acceptable Δε as described above, we are interested in investigation of the application of them as the compositions in blue phase liquid crystals (BPLC) to possibly increase the Kerr constant, which is key parameter for practical BPLC. After detailed screening the composition and contents, we got a LC mixture containing BPEBs **1**, **3–4**, **8–9**, **14**, **15**, **21** (5 wt % each) and other liquid crystal mixture (cp: 83.0 °C; Δn = 0.230, and Δε = 29.6, at 25 °C), which shows the properties of Δn = 0.283, and Δε = 29.0 at 25 °C, which is expected to have high potential applications as BPLC [24]. After adding chiral dopants (R811: 10 wt % and BDH1281: 7 wt %), we obtained a BPLC with a blue phase temperature range of 8 K (from 41 to 33 °C, on the second cooling run). Figure 4 shows a typical BP texture of the obtained BPLC at 36 °C.

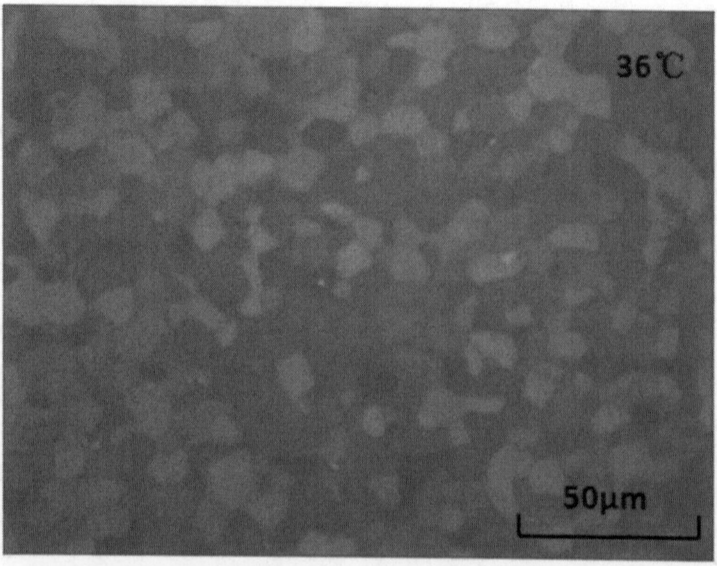

Figure 4. The typical blue phase texture.

3. EXPERIMENTAL SECTION

3.1. General Method

All organic starting materials and catalysts are analytically pure and used without further purification. Nuclear magnetic resonance (NMR) spectra were recorded on a JEOL ECA-300 spectrometer (Tokyo, Japan) using $CDCl_3$ as solvent at 298 K. ^{1}H-NMR (300 MHz) chemical shifts (δ) were referenced to internal standard TMS (for 1H, δ = 0.00 ppm). ^{13}C-NMR (75 MHz) chemical shifts were referenced to internal solvent $CDCl_3$ (for ^{13}C, δ = 77.16 ppm). Mass spectra (MS) were obtained on a Shimadzu GCMS-QP2010S (Kyoto, Japan). Element analyses were obtained with a Flash EA 1112 Element Analyzer (Thermo Fisher Scientific, Waltham, MA, USA). Polarizing microscope LWT300LPT (CEWEI photoelectric technology Co. Ltd., Shanghai, China) equipped with a Weitu WT-3000 hot stage and a TCA 5.0 MP camera was used to observe and record the optical textures of liquid crystal samples. The calorimetric studies were performed on a TA Instruments DSC 2010 (TA Instruments, New Castle, DE, USA) with a heating temperature rate of 10 °C/min. A NAR-4T Abbe refractometer (ATAGO Co. Ltd., Tokyo, Japan) was used to measure optical anisotropy (Δn), and a 3522-50 LCR Hitester (HIOKI E.E. Co., Ueda, Japan) for dielectric anisotropy (Δε). All the samples for measuring the Δn and Δε were composed of BPEB and nematic host LC at a ratio of 5–10/95–90 (wt %/wt %). The nematic host LC (SLC960524) was prepared by our laboratory, which has the values of Δn = 0.1202 (589 nm) and Δε = 3.121 (1000 Hz) at 25 °C.

3.2. A Typical Experimental Procedure for Synthesis of BPEB 12 and the Characterization Data of All the BPEBs

As shown in Scheme 2, BPEBs and analogues (**1–25**) were synthesized by the similar synthetic route, and their structures were characterized by ^{1}H-NMR, ^{13}C-NMR (for BPEBs **6** and **7**, the ^{13}C-NMR could not be obtained due to their very low solubility in $CDCl_3$, DMSO-d_6 or DMF-d_7), and elemental analyses. In this section, the synthetic procedure of BPEB **12** was only described in details, and the characterization data of all the other BPEBs are given.

Preparation of 2,6-difluoro-4-n-propylphenyl acetylene (12c) (See Scheme S1).

2-Methyl-3-butyn-2-ol (21.8 g, 0.3 mol) was added to a mixture of 2,6-difluoro-4-n-propyl-1- iodobenzene (56.4 g, 0.2 mol), tetrakis(triphenylphosphine) palladium(0) (1.0 g, 0.87 mmol), CuBr (0.5 g) and LiBr (2.0 g) in triethylamine (50 mL) with stirring at room temperature. After the mixture was heated at 60 °C for 5 h, it was then cooled to room temperature and the saturated NH_4Cl aqueous solution (100 mL) and ethyl acetate (200 mL) were added. After separation of the organic phase, the aqueous phase was extracted by ethyl acetate (3 × 150 mL), and the combined organic extracts are dried by K_2CO_3 and concentrated on a rotary evaporator, the intermediate **12b** was obtained by column chromatographic separation (40.5 g, 0.17 mol, 85.0%) for the next reaction.

Sodium hydroxide (20.0 g, 0.5 mol) was added to a solution of **12b** (40.0 g) in toluene (150 mL), and then the obtained mixture was heated under reflux for 5 h. After removal of the insoluble excess of Sodium hydroxide by filtration and the solvent under reduced pressure, **12c** was isolated by column chromatography as light yellow oil (17.2 g, 0.095 mol, 55.9%). Characterization data for **12c**: ^{1}H-NMR (300 MHz, $CDCl_3$) δ 6.70 (d, 2H, $^{3}J_{F\text{-}C\text{-}C\text{-}H}$ = 8.4 Hz), 6.67 (s, 1H), 3.44 (s, 1H), 2.52 (t, 2H, J = 7.5 Hz), 1.61–1.54 (m, 2H), 0.89 (t, 3H, J = 7.5 Hz); ^{13}C-NMR (75 MHz, $CDCl_3$) δ 163.5 (dd, $^{1}J_{C\text{-}F}$ = 253.6 Hz, $^{5}J_{C\text{-}F}$ = 6.0 Hz), 147.1 (t, $^{3}J_{C\text{-}F}$ = 9.0 Hz), 111.1 (dd, $^{2}J_{C\text{-}F}$ = 17.1 Hz, $^{4}J_{C\text{-}F}$ = 6.0 Hz), 98.2 (t, $^{2}J_{C\text{-}F}$ = 19.8 Hz), 86.5, 70.9, 37.8, 23.7, 13.4; MS m/z (% rel. intensity) 180 (M^+, 44), 151 (100); Anal. calcd for $C_{11}H_{10}F_2$: C, 73.33; H, 5.56. Found: C, 73.62; H, 5.59.

Preparation of [2,6-difluoro-4-(2′,6′-difluoro-4′-n-propylphenyl) ethynyl] phenyl acetylene (12g) (See Scheme S2).

A mixture of 2-bromo-1,3-difluoro-5-iodobenzene (11.16 g, 0.035 mol), ethynyl trimethylsilane (3.8 g, 0.039 mol), tetrakis(triphenylphosphine) palladium(0) (0.5 g, 0.44 mmol) and CuI (1.0 g) in triethylamine (40.0 mL) was stirred in under argon at room temperature for 16 h. After work-up as described for **12b**, **12e** was obtained in 85.7% (10.0 g, 0.03 mol).

A mixture of **12e** (5.6 g), **12c** (3.0 g, 0.017 mol), tetrakis(triphenylphosphine)palladium(0) (0.2 g, 0.17 mmol) and CuI (0.4 g) in triethylamine (40.0 mL) was heated with strring at 60 °C for 5 h, After work-up as described for **12b**, the intermediated **12f** was obtained as colorless solid in 76.4% (4.8 g, 0.013 mol). And then stirring a mixture of **12f** (4.8 g, 0.013 mol) and K_2CO_3 (0.1 g, 0.7 mmol) in methanol (150 mL) at room temperature for 5 h, after work up as described for the isolation of **12c**, **12g** was isolated as colorless solid in 84.6% (3.47 g, 0.011 mol). Characterization data for **12g**: ^{1}H-NMR (300 MHz, $CDCl_3$) δ 7.11 (d, 2H, $^3J_{F\text{-}C\text{-}C\text{-}H}$ = 7.5 Hz), 6.77 (d, 2H, $^3J_{F\text{-}C\text{-}C\text{-}H}$ = 8.1 Hz), 3.59 (s, 1H), 2.59 (t, 2H, J = 7.5 Hz), 1.68–1.60 (m, 2H), 0.94 (t, 3H, J = 7.4 Hz); ^{13}C-NMR (300 MHz, $CDCl_3$) δ 163.3 (dd, $^1J_{C\text{-}F}$ = 255.0 Hz, $^5J_{C\text{-}F}$ = 6.6 Hz), 162.8 (dd, $^1J_{C\text{-}F}$ = 254.2 Hz, $^5J_{C\text{-}F}$ = 5.8 Hz), 147.6 (t, $^3J_{C\text{-}F}$ = 9.1 Hz), 125.1 (t, $^3J_{C\text{-}F}$ = 11.9 Hz), 114.5 (dd, $^2J_{C\text{-}F}$ = 18.2 Hz, $^4J_{C\text{-}F}$ = 6.8.Hz), 111.4 (dd, $^2J_{C\text{-}F}$ = 17.3 Hz, $^4J_{C\text{-}F}$ = 6.3.Hz), 102.1 (t, $^2J_{C\text{-}F}$ = 19.6 Hz), 98.4 (t, $^2J_{C\text{-}F}$ = 19.6 Hz), 88.8, 88.9, 80.7, 70.5, 38.0, 23.8, 13.6; MS m/z (% rel. intensity) 316 (M^+, 56); 287 (100); Anal. calcd for $C_{19}H_{12}F_4$: C, 72.15; H, 3.80. Found: C, 72.52; H, 3.98.

Preparation of 1-[(2′,6′-difluoro-4′-n-propylphenyl)ethynyl]-4-[(3″,5″-difluoro-4″-trifluoro-methoxy-phenyl) ethynyl]-3,5-difluorobenzene (BPEB 12) (See Scheme S3).

A solution of 4-bromo-2,6-difluoro(trifluoromethoxy)benzene (2.2 g, 8.04 mmol) in toluene (10.0 mL) was added into a solution of **12g** (1.69 g, 5.36 mmol), tetrakis(triphenylphosphine)palladium(0) (0.1 g, 0.09 mmol), and CuI (0.2 g) in triethylamine (20.0 mL) under argon at 60 °C, and then the obtained mixture was stirred at 60 °C for 5 h. After work-up as described for **12c**, the desired **BPEB 12** was isolated as colorless solid in 79.1% (2.17 g, 4.24 mmol). Characterization data for **BPEB 12**: ^{1}H-NMR (300 MHz, $CDCl_3$) δ 7.24 (d, 2H, $^3J_{F\text{-}C\text{-}C\text{-}H}$ = 7.5), 7.15 (d, 2H, $^3J_{F\text{-}C\text{-}C\text{-}H}$ = 7.5 Hz), 6.78 (d, 2H, $^3J_{F\text{-}C\text{-}C\text{-}H}$ = 8.4 Hz), 2.60 (t, 2H, J = 7.5 Hz), 1.69–1.61 (m, 2H), 0.95 (t, 3H, J = 7.4 Hz); ^{13}C-NMR (75 MHz, $CDCl_3$) δ 162.9 (dd, $^1J_{C\text{-}F}$ = 254.3 Hz, $^5J_{C\text{-}F}$ = 5.7 Hz), 162.6 (dd, $^1J_{C\text{-}F}$ = 254.8 Hz, $^5J_{C\text{-}F}$ = 6.0 Hz), 155.9 (dd, $^1J_{C\text{-}F}$ = 253.9 Hz, $^5J_{C\text{-}F}$ = 3.6 Hz), 147.7 (t, $^3J_{C\text{-}F}$ = 9.0 Hz), 126.3 (t, $^3J_{C\text{-}F}$ = 15.6 Hz), 125.6 (t, $^3J_{C\text{-}F}$ = 12.1 Hz), 123.0 (t, $^3J_{C\text{-}F}$ = 10.6 Hz), 120.5 (q, $^1J_{C\text{-}F}$ = 260.3 Hz), 116.0 (dd, $^2J_{C\text{-}F}$ = 16.2 Hz, $^4J_{C\text{-}F}$ = 5.0 Hz), 114.6 (dd, $^2J_{C\text{-}F}$ = 18.2 Hz, $^4J_{C\text{-}F}$ = 7.0 Hz), 111.4 (dd, $^2J_{C\text{-}F}$ = 17.2 Hz, $^4J_{C\text{-}F}$ = 5.7 Hz), 102.0 (t, $^2J_{C\text{-}F}$ = 18.7 Hz), 98.5 (t, $^2J_{C\text{-}F}$ = 19.7 Hz), 96.8, 95.5, 81.1, 79.2, 38.0, 23.8, 13.6; Anal. calcd for $C_{26}H_{13}F_9O$: C, 60.93; H, 2.54. Found: C, 61.33; H, 2.50.

Characterization data for **BPEB 1**: ^{1}H-NMR (300 MHz, $CDCl_3$) δ 7.56–7.45 (m, 6H), 6.87 (d, 2H, J = 8.8 Hz), 6.78 (d, 2H, J = 8.2 Hz), 4.04 (q, 2H, J = 7.0 Hz), 2.65 (q, 2H, J = 7.6 Hz), 1.42 (t, 3H, J = 7.0 Hz), 1.24 (t, 3H, J = 7.6 Hz); ^{13}C-NMR (75 MHz, $CDCl_3$) δ 162.8 (dd, J = 253.1 Hz, J = 6.0 Hz), 159.3, 147.9 (t, J = 8.9 Hz), 133.2, 131.7, 131.4, 124.1, 122.2, 115.0, 114.7, 110.7 (dd, J = 16.9

Hz, J = 6.3 Hz), 99.4 (t, J = 19.9 Hz), 98.1, 91.8, 87.9, 78.9, 63.6, 28.9, 14.9, 14.8; Anal. calcd for $C_{26}H_{20}F_2O$: C, 80.83; H, 5.18. Found: C, 81.03; H, 4.90.

Characterization data for **BPEB 2**: ^{1}H-NMR (300 MHz, $CDCl_3$) δ 7.56–7.45 (m, 6H), 6.87 (d, 2H, J = 8.7 Hz), 6.76 (d, 2H, J = 8.0 Hz), 4.04 (q, 2H, J = 7.0 Hz), 2.58 (t, 2H, J = 7.4 Hz), 1.70–1.58 (m, 2H), 1.42 (t, 3H, J = 7.0 Hz), 0.95 (t, 3H, J = 7.3 Hz); ^{13}C-NMR (75 MHz, $CDCl_3$) δ 162.8 (dd, J = 253.0 Hz, J = 6.0 Hz), 159.3, 146.4 (t, J = 9.0 Hz), 133.2, 131.7, 131.4, 124.1, 122.2, 115.0, 114.7, 111.3 (dd, J = 16.6 Hz, J = 6.0 Hz), 99.4 (t, J = 20.1 Hz), 98.1, 91.8, 87.9, 78.3, 63.6, 38.0, 23.9, 14.8, 13.7; Anal. calcd for $C_{27}H_{22}F_2O$: C, 81.00; H, 5.50. Found: C, 81.43; H, 5.74.

Characterization data for **BPEB 3**: ^{1}H-NMR (300 MHz, $CDCl_3$) δ 7.56–7.45 (m, 6H), 6.87 (d, 2H, J = 8.7 Hz), 6.76 (d, 2H, J = 8.1 Hz), 4.04 (q, 2H, J = 7.0 Hz), 2.60 (t, 2H, J = 7.5 Hz), 1.61–1.54 (m, 2H), 1.45–1.30 (m, 5H), 0.94 (t, 3H, J = 7.3 Hz); ^{13}C-NMR (75 MHz, $CDCl_3$) δ 162.8 (dd, J = 253.0 Hz, J = 6.1 Hz), 159.3, 146.7 (t, J = 9.0 Hz), 133.2, 131.7, 131.4, 124.1, 122.2, 115.0, 114.7, 111.2 (dd, J = 16.8 Hz, J = 6.0 Hz), 99.4 (t, J = 20.0 Hz), 98.1, 91.8, 87.9, 78.3, 63.6, 35.7, 32.8, 22.2, 14.8, 13.9; Anal. calcd for $C_{28}H_{24}F_2O$: C, 81.16; H, 5.80. Found: C, 81.53; H, 5.84.

Characterization data for **BPEB 4**: ^{1}H-NMR (300 MHz, $CDCl_3$) δ 7.55–7.45 (m, 6H), 6.87 (d, 2H, J = 8.7 Hz), 6.76 (d, 2H, J = 8.1 Hz), 4.04 (q, 2H, J = 7.0 Hz), 2.60 (t, 2H, J = 7.5 Hz), 1.66–1.57 (m, 2H), 1.43 (t, 3H, J = 7.0 Hz), 1.36–1.25 (m, 4H), 0.91 (t, 3H, J = 6.9 Hz); ^{13}C-NMR (75 MHz, $CDCl_3$) δ 162.8 (dd, J = 253.2 Hz, J = 6.1 Hz), 159.3, 146.7 (t, J = 9.0 Hz), 133.2, 131.7, 131.4, 124.1, 122.2, 115.0, 114.6, 111.2 (dd, J = 17.0 Hz, J = 6.1 Hz), 99.4 (t, J = 19.5 Hz), 98.1, 91.8, 87.9, 78.3, 63.6, 35.9, 31.4, 30.4, 22.5, 14.8, 14.1; Anal. calcd for $C_{29}H_{26}F_2O$: C, 81.31; H, 6.07. Found: C, 81.26; H, 6.34.

Characterization data for **BPEB 5**: ^{1}H-NMR (300 MHz, $CDCl_3$) δ 7.54–7.39 (m, 7H), 6.94 (d, 2H, J = 8.4 Hz), 6.87 (d, 2H, J = 8.3 Hz), 4.04 (q, 2H, J = 6.7 Hz), 2.60 (t, 3H, J = 7.3 Hz), 1.72–1.59 (m, 2H), 1.43 (t, 3H, J = 6.8 Hz), 0.96 (t, 3H, J = 7.2 Hz); ^{13}C-NMR (75 MHz, $CDCl_3$) δ 162.7 (d, J = 251.6 Hz), 159.3, 146.1 (d, J = 7.2 Hz), 133.2, 133.1, 131.6, 131.4, 124.3, 123.8, 122.6, 115.5 (d, J = 20.3 Hz), 114.9 (d, J = 27.2 Hz), 114.7, 108.9 (d, J = 16.0 Hz), 93.6, 91.6, 88.0, 84.8, 63.6, 37.9, 24.1, 14.9, 13.8; Anal. calcd for $C_{27}H_{23}FO$: C, 84.82; H, 6.02. Found: C, 84.88; H, 6.13.

Characterization data for **BPEB 6**: ^{1}H-NMR (300 MHz, $CDCl_3$) δ 7.64–7.53 (m, 8H), 7.49–7.45 (m, 2H), 7.41 (d, 1H, J = 7.5 Hz), 6.96–6.92 (m, 2H), 6.88 (d, 2H, J = 8.7 Hz), 4.06 (q, 2H, J = 7.0 Hz), 2.61 (t, 2H, J = 7.3 Hz), 1.72–1.60 (m, 2H), 1.43 (t, 3H, J = 7.0 Hz), 0.95 (t, 3H, J = 7.3 Hz); Anal. calcd for $C_{33}H_{27}FO$: C, 86.46; H, 5.90. Found: C, 86.98; H, 5.93.

Characterization data for **BPEB 7**: ^{1}H-NMR (300 MHz, $CDCl_3$) δ 7.66–7.55 (m, 8H), 7.48 (d, 2H, J = 8.8 Hz), 6.88 (d, 2H, J = 8.8 Hz), 6.77 (d, 2H, J = 8.1 Hz), 4.06 (q, 2H, J = 7.0 Hz), 2.59 (t, 2H, J = 7.4 Hz), 1.71–1.59 (m, 2H), 1.43 (t, 3H, J = 7.0 Hz), 0.95 (t, 3H, J = 7.4 Hz); Anal. calcd for $C_{33}H_{26}F_2O$: C, 83.19; H, 5.46. Found: C, 83.26; H, 5.72.

Characterization data for **BPEB 8**: ^{1}H-NMR (300 MHz, $CDCl_3$) δ 7.58–7.48 (m, 4H), 7.17 (d, 2H, J = 7.8 Hz), 6.77 (d, 2H, J = 8.1 Hz), 2.59 (t, 2H, J = 7.4 Hz),1.71–1.59 (m, 2H), 0.95 (t, 3H, J = 7.3 Hz); ^{13}C-NMR (75 MHz, $CDCl_3$) δ 162.8 (dd, J = 253.3 Hz, J = 5.8 Hz), 155.8 (dd, J = 255.8 Hz, J = 3.6 Hz), 146.8 (t, J =8.9 Hz), 131.8, 131.7, 125.6 (t, J = 17.2 Hz), 123.8 (t, J = 9.1 Hz), 122.1, 120.6 (q, J = 259.6 Hz), 115.8 (d, J = 17.5 Hz, J = 6.2 Hz), 111.3 (dd, J = 16.9 Hz, J = 6.1 Hz), 99.2 (t, J = 19.9 Hz), 97.6, 92.0, 87.9, 79.1, 38.0, 23.8, 13.6; Anal. calcd for $C_{26}H_{15}F_7O$: C, 65.55; H, 3.15. Found: C, 64.98; H, 3.13.

Characterization data for **BPEB 9**: ^{1}H-NMR (300 MHz, $CDCl_3$) δ 7.57–7.47 (m, 4H), 7.16–7.11 (m, 2H), 6.76 (d, 2H, J = 8.1 Hz), 2.59 (t, 2H, J = 7.4 Hz), 1.71–1.58 (m, 2H), 0.95 (t, 3H, J = 7.3 Hz); ^{13}C-NMR (75 MHz, $CDCl_3$) δ 162.8 (dd, J = 253.1 Hz, J = 5.8 Hz), 151.1 (ddd, J = 251.1 Hz, J = 10.4 Hz, J = 4.9 Hz), 146.7 (t, J = 9.0 Hz), 140.5 (dt, J = 255.7 Hz, J = 15.3 Hz), 131.8, 131.7, 123.5, 122.4, 119.0 9 (td), 116.0 (ddd, J = 15.2 Hz, J = 7.3 Hz), 111.3 (dd, J =17.0 Hz, J = 5.9 Hz), 99.2 (t, J = 20.0 Hz), 97.7, 90.6, 88.2, 78.9, 38.0, 23.9, 13.6; Anal. calcd for $C_{25}H_{15}F_5$: C, 73.17; H, 3.66. Found: C, 73.02; H, 3.51.

Characterization data for **BPEB 10**: ^{1}H-NMR (300 MHz, $CDCl_3$) δ 7.47 (t, 1H, J = 7.6 Hz), 7.35–7.30 (m, 2H), 7.20 (d, 2H, J = 7.8 Hz), 6.77 (d, 2H, J = 8.2 Hz), 2.59 (t, 2H, J = 7.4 Hz), 1.71–1.59 (m, 2H), 0.95 (t, 3H, J = 7.3 Hz); ^{13}C-NMR (75 MHz, $CDCl_3$) δ 162.8 (dd, J = 253.9 Hz, J = 6.0 Hz), 162.3 (d, J = 253.5 Hz), 155.8 (dd, J = 255.4 Hz, J = 3.8 Hz), 147.2 (t, J = 9.1 Hz), 133.3, 127.6 (d, J = 3.2 Hz), 126.1 (t, J = 15.8 Hz), 125.6 (d, J = 9.3 Hz), 123.3 (t, J = 11.0 Hz), 120.5 (q, J = 261.0 Hz), 118.7 (d, J = 22.7 Hz), 115.9 (dd, J = 15.5 Hz, J = 4.3 Hz), 111.4 (dd, J = 17.5 Hz, J = 5.6 Hz), 111.1, 98.8 (t, J = 19.6 Hz), 96.4, 92.6, 85.5, 80.0, 40.0, 23.8, 13.6; Anal. calcd for $C_{26}H_{14}F_8O$: C, 63.16; H, 2.83. Found: C, 63.03; H, 2.56.

Characterization data for **BPEB 11**: ^{1}H-NMR (300 MHz, $CDCl_3$) δ 7.45 (t, 1H, J = 7.4 Hz), 7.36–7.26 (m, 2H), 7.21–7.14 (m, 2H), 6.77 (d, 2H, J = 8.2 Hz), 2.59 (t, 2H, J = 7.3 Hz), 1.71–1.58 (m, 2H), 0.95 (t, 3H, J = 7.3 Hz); ^{13}C-NMR (75 MHz, $CDCl_3$) δ 162.8 (dd, J = 253.6 Hz, J = 5.6 Hz), 162.3 (d, J = 253.2 Hz), 151.1 (ddd, J = 251.0 Hz, J = 10.5 Hz, J = 4.8 Hz), 147.2 (t, J = 8.9 Hz), 140.6 (dt, J = 250.7 Hz, J = 10.9 Hz), 133.3, 127.6 (d, J = 3.3 Hz), 125.3 (d, J = 9.3 Hz), 118.6 (dd, J = 12.4 Hz, J = 6.9 Hz), 116.1 (dd, J = 15.5 Hz, J = 7.4 Hz), 111.4, 111.4 (dd, J = 17.8 Hz, J = 6.3 Hz), 98.9 (t, J = 19.4 Hz), 96.5 (d, J = 3.2 Hz), 92.9, 84.1, 79.9, 38.0, 23.8, 13.6; Anal. calcd for $C_{25}H_{14}F_6$: C, 70.09; H, 3.27. Found: C, 70.33; H, 3.46.

Characterization data for **BPEB 13**: ^{1}H-NMR (300 MHz, $CDCl_3$) δ 7.22–7.12 (m, 4H), 6.78 (d, 2H, J = 8.2 Hz), 2.60 (t, 2H, J = 7.4 Hz), 1.71–1.59 (m, 2H), 0.95 (t, 3H, J = 7.3 Hz); $_{13}$C-NMR (75 MHz, $CDCl_3$) δ 162.8 (dd, J = 254.0 Hz, J = 5.4 Hz), 162.6 (dd, J = 255.1 Hz, J = 6.4 Hz), 151.2 (ddd, J = 252.3 Hz, J = 10.7 Hz, J = 4.4 Hz), 147.6 (t, J = 6.4 Hz), 140.9 (dt, J = 256.0 Hz, J = 15.4 Hz), 125.3 (t, J = 12.2 Hz), 118.4, 116.2 (dd, J = 15.5 Hz, J = 7.6 Hz), 114.5 (dd, J =17.7 Hz, J = 7.1 Hz), 111.4 (dd, J = 16.4 Hz, J = 5.1 Hz), 102.3 (t, J = 21.5 Hz), 98.6 (t, J = 20.2 Hz), 97.3, 95.6, 80.9, 77.8, 38.0, 23.7, 13.5; Anal. calcd for $C_{25}H_{13}F_7$: C, 67.26; H, 2.91. Found: C, 69.52; H, 2.66.

Characterization data for **BPEB 14**: ^{1}H-NMR (300 MHz, $CDCl_3$) δ 7.57–7.48 (m, 4H), 7.17 (d, 2H, J = 7.9 Hz), 6.77 (d, 2H, J = 8.1 Hz), 2.61 (t, 2H, J = 7.5 Hz), 1.65–1.55 (m, 2H), 1.42–1.30 (m, 2H), 0.94 (t, 3H, J = 7.3 Hz); ^{13}C-NMR (75 MHz, $CDCl_3$) δ 162.8 (dd, J = 253.4 Hz, J = 6.2 Hz), 155.8 (dd, J = 255.3 Hz, J = 3.7 Hz), 147.0 (t, J = 9.1 Hz), 131.8, 131.7, 125.8 (t, J = 16.3 Hz), 123.8 (t, J = 8.8 Hz), 122.1, 120.4 (q, J = 258.9 Hz), 115.8 (dd, J = 17.5 Hz, J = 6.2 Hz), 111.3 (dd, J = 16.7 Hz, J = 5.7 Hz), 99.2 (t, J = 20.1 Hz), 97.6, 92.0, 87.9, 79.0, 35.7, 32.8, 22.3, 13.9; Anal. calcd for $C_{27}H_{17}F_7O$: C, 66.12; H, 3.47. Found: C, 66.01; H, 3.36.

Characterization data for **BPEB 15**: ^{1}H-NMR (300 MHz, $CDCl_3$) δ 7.56–7.46 (m, 4H), 7.13 (t, 2H, J = 6.9 Hz), 6.76 (d, 2H, J = 8.1 Hz), 2.60 (t, 2H, J = 7.4 Hz), 1.64–1.54 (m, 2H), 1.42–1.30 (m, 2H), 0.94 (t, 3H, J = 7.3 Hz); ^{13}C-NMR (75 MHz, $CDCl_3$) δ 162.8 (dd, J = 253.4 Hz, J = 6.0 Hz), 151.1 (ddd, J = 250.6 Hz, J = 10.1 Hz, J = 4.2 Hz), 146.9 (t, J = 9.0 Hz), 140.4 (dt, J = 254.7 Hz, J = 14.9 Hz), 131.8, 131.7, 123.5, 122.4, 119.1 (td), 116.0 (ddd, J = 15.4 Hz, J = 7.3 Hz, J = 4.8 Hz), 111.2 (dd, J =17.1 Hz, J = 5.9 Hz), 99.2 (t, J = 19.8 Hz), 97.7, 90.6, 88.2, 78.9, 35.7, 32.8, 22.3, 13.9; Anal. calcd for $C_{26}H_{17}F_5$: C, 73.58; H, 4.01. Found: C, 73.51; H, 3.96.

Characterization data for **BPEB 16**: ^{1}H-NMR (300 MHz, $CDCl_3$) δ 7.47 (t, 1H, J = 7.5 Hz), 7.39–7.29 (m, 2H), 7.20 (d, 2H, J = 7.7 Hz), 6.77 (d, 2H, J = 8.2 Hz), 2.61 (t, 2H, J = 7.6 Hz), 1.65–1.55 (m, 2H), 1.42–1.30 (m, 2H), 0.94 (t, 3H, J = 7.3 Hz); ^{13}C-NMR (75 MHz, $CDCl_3$) δ 162.8 (dd, J = 253.8 Hz, J = 5.8 Hz), 162.3 (d, J = 253.7 Hz), 155.8 (dd, J = 255.5 Hz, J = 3.7 Hz), 147.5 (t, J = 8.9 Hz), 133.3, 127.6 (d, J = 3.4 Hz), 126.1 (t, J = 15.8 Hz), 125.6 (d, J = 9.4 Hz), 123.3 (t, J = 11.2 Hz), 120.5 (q, J = 261.0 Hz), 118.7 (d, J = 22.8 Hz), 115.9 (dd, J = 16.2 Hz, J = 6.4 Hz), 111.3 (dd, J = 16.1 Hz, J = 5.8 Hz), 111.1, 98.8 (t, J = 19.5 Hz), 96.4, 92.5, 85.5, 80.0, 35.7, 32.8, 22.3, 13.9; Anal. calcd for $C_{27}H_{16}F_8O$: C, 63.78; H, 3.15. Found: C, 63.61; H, 3.26.

Characterization data for **BPEB 17**: ^{1}H-NMR (300 MHz, $CDCl_3$) δ 7.44 (t, 1H, J = 7.4 Hz), 7.33–7.27 (m, 2H), 7.16 (t, 2H, J = 6.6 Hz), 6.76 (d, 2H, J = 8.2 Hz), 2.60 (t, 2H, J = 7.5 Hz), 1.64–1.54 (m, 2H), 1.42–1.29 (m, 2H), 0.94 (t, 3H, J = 7.3 Hz); ^{13}C-NMR (75 MHz, $CDCl_3$) δ 162.8 (dd, J = 253.7 Hz, J = 5.8 Hz), 162.3 (d, J = 253.4 Hz), 151.1 (ddd, J = 250.9 Hz, J = 10.4 Hz, J = 4.3 Hz), 147.4 (t, J = 9.0 Hz), 140.7 (dt, J = 255.1 Hz, J = 14.7 Hz), 133.3, 127.6 (d, J = 3.3 Hz), 125.3 (d, J = 9.5 Hz), 118.6 (dd, J = 14.2 Hz, J = 9.1 Hz), 116.1 (dd, J = 15.4 Hz, J = 7.3 Hz), 111.5, 111.3 (dd, J = 17.0 Hz, J = 5.9 Hz), 98.8 (t, J = 19.6 Hz), 96.5 (d, J = 3.1 Hz), 92.9, 84.1, 79.9, 35.7, 32.8, 22.3, 13.9; Anal. calcd for $C_{26}H_{16}F_6$: C, 70.59; H, 3.62. Found: C, 70.68; H, 3.36.

Characterization data for **BPEB 18**: ^{1}H-NMR (300 MHz, $CDCl_3$) δ 7.23 (d, 2H, J = 7.7 Hz), 7.14 (d, 2H, J = 7.3 Hz), 6.78 (d, 2H, J = 8.2 Hz), 2.62 (t, 2H, J = 7.6 Hz), 1.65–1.55 (m, 2H), 1.41–1.30 (m, 2H), 0.94 (t, 3H, J = 7.3 Hz); ^{13}C-NMR (75 MHz, $CDCl_3$) δ 162.9 (dd, J = 254.1 Hz, J = 5.8 Hz), 162.7 (dd, J = 255.4 Hz, J = 6.2 Hz), 155.9 (dd, J = 255.8 Hz, J = 3.6 Hz), 147.9 (t, J = 9.0 Hz), 126.3 (t, J = 15.8 Hz), 125.6 (t, J = 12.0 Hz), 123.0 (t, J = 10.1 Hz), 120.5 (q, J = 260.8 Hz), 116.0 (dd, J = 17.0 Hz, J = 6.6 Hz), 114.6 (dd, J = 17.8 Hz, J = 7.9

Hz), 111.3 (dd, J = 17.0 Hz, J = 5.6 Hz), 102.0 (t, J = 18.5 Hz), 98.2 (t, J = 19.9 Hz), 96.9, 95.5, 81.1, 79.2, 35.7, 32.7, 22.2, 13.8; Anal. calcd for $C_{27}H_{15}F_9O$: C, 61.60; H, 2.85. Found: C, 60.99; H, 2.64.

Characterization data for **BPEB 19**: ^{1}H-NMR (300 MHz, $CDCl_3$) δ 7.20 (t, 2H, J = 6.7 Hz), 7.12 (d, 2H, J = 7.1 Hz), 6.77 (d, 2H, J = 8.2 Hz), 2.62 (t, 2H, J = 7.5 Hz), 1.65–1.55 (m, 2H), 1.42–1.30 (m, 2H), 0.94 (t, 3H, J = 7.3 Hz); ^{13}C-NMR (75 MHz, $CDCl_3$) δ 162.8 (dd, J = 254.1 Hz, J = 5.7 Hz), 162.6 (dd, J = 254.7 Hz, J = 6.2 Hz), 151.1 (ddd, J = 251.3 Hz, J = 10.4 Hz, J = 4.4 Hz), 147.9 (t, J = 9.2 Hz), 140.9 (dt, J = 256.0 Hz, J = 15.6 Hz), 125.2 (t, J = 12.0 Hz), 118.4 (td), 116.3 (ddd, J = 15.4 Hz, J = 7.3 Hz), 114.6 (dd, J =18.2 Hz, J = 7.6 Hz), 111.4 (dd, J = 17.2 Hz, J = 5.7 Hz), 102.2 (t, J = 19.8 Hz), 98.4 (t, J = 20.1 Hz), 97.3, 95.6, 80.9, 77.8, 35.7, 32.8, 22.3, 13.9; Anal. calcd for $C_{26}H_{15}F_7$: C, 67.83; H, 3.26. Found: C, 67.77; H, 3.44.

Characterization data for **BPEB 20**: ^{1}H-NMR (300 MHz, $CDCl_3$) δ 7.57–7.48 (m, 4H), 7.16 (d, 2H, J = 7.7 Hz), 6.77 (d, 2H, J = 8.2 Hz), 2.60 (t, 2H, J = 7.5 Hz), 1.65–1.56 (m, 2H), 1.38–1.28 (m, 4H), 0.91 (t, 3H, J = 6.7 Hz); ^{13}C-NMR (75 MHz, $CDCl_3$) δ 162.8 (dd, J = 253.4 Hz, J = 6.0 Hz), 155.8 (dd, J = 255.4 Hz, J = 3.7 Hz), 147.0 (t, J = 9.1 Hz), 131.8, 131.7, 125.8 (t, J = 16.5 Hz), 123.8 (t, J = 8.7 Hz), 122.2, 120.6 (q, J = 259.6 Hz), 115.8 (dd, J = 17.6 Hz, J = 6.4 Hz), 111.2 (dd, J = 17.1 Hz, J = 6.1 Hz), 99.1 (t, J = 19.9 Hz), 97.6, 92.0, 87.9, 79.1, 35.9, 31.3, 30.3, 22.5, 14.0; Anal. calcd for $C_{28}H_{19}F_7O$: C, 66.67; H, 3.77. Found: C, 66.35; H, 3.46.

Characterization data for **BPEB 21**: ^{1}H-NMR (300 MHz, $CDCl_3$) δ 7.56–7.46 (m, 4H), 7.16–7.11 (m, 2H), 6.76 (d, 2H, J = 8.2 Hz), 2.60 (t, 2H, J = 7.6 Hz), 1.66–1.56 (m, 2H), 1.38–1.29 (m, 4H), 0.91 (t, 3H, J = 6.6 Hz); ^{13}C-NMR (75 MHz, $CDCl_3$) δ 162.8 (dd, J = 253.2 Hz, J = 6.0 Hz), 151.1 (ddd, J = 250.7 Hz, J = 10.3 Hz, J = 4.5 Hz), 147.0 (t, J = 8.9 Hz), 140.5 (dt, J = 256.0 Hz, J = 15.6 Hz), 131.8, 131.7, 123.5, 122.4, 119.1 (td), 116.0 (ddd, J = 15.2 Hz, J = 7.2 Hz, J = 4.3 Hz), 111.2 (dd, J = 16.6 Hz, J = 5.6 Hz), 99.2 (t, J = 19.7 Hz), 97.7, 90.6, 88.2, 78.9, 35.9, 31.3, 30.4, 22.5, 14.0; Anal. calcd for $C_{27}H_{19}F_5$: C, 73.97; H, 4.34. Found: C, 73.49; H, 4.44.

Characterization data for **BPEB 22**: ^{1}H-NMR (300 MHz, $CDCl_3$) δ 7.46 (t, 1H, J = 7.5 Hz), 7.34–7.29 (m, 2H), 7.19 (d, 2H, J = 7.1 Hz), 6.77 (d, 2H, J = 8.2 Hz), 2.60 (t, 2H, J = 7.5 Hz), 1.61 (m, 2H), 1.36–1.30 (m, 4H), 0.91 (t, 3H, J = 6.6 Hz); ^{13}C-NMR (75 MHz, $CDCl_3$) δ 162.8 (dd, J = 253.9 Hz, J = 5.8 Hz), 162.3 (d, J = 253.6 Hz), 155.8 (dd, J = 255.5 Hz, J = 3.5 Hz), 147.5 (t, J = 9.0 Hz), 133.3, 127.6 (d, J = 3.4 Hz), 126.1 (t, J = 15.8 Hz), 125.6 (d, J = 9.5 Hz), 123.3 (t, J = 10.9 Hz), 120.5 (q, J = 261.0 Hz), 118.7 (d, J = 22.7 Hz), 115.9 (dd, J = 16.0 Hz, J = 6.5 Hz), 111.3 (dd, J = 14.6 Hz, J = 6.1 Hz), 111.1, 98.8 (t, J = 19.9 Hz), 96.4, 92.6, 85.5, 80.0, 36.0, 31.3, 30.3, 22.5, 14.0; Anal. calcd for $C_{28}H_{18}F_8O$: C, 64.37; H, 3.45. Found: C, 63.89; H, 3.26.

Characterization data for **BPEB 23**: ^{1}H-NMR (300 MHz, $CDCl_3$) δ 7.44 (t, 1H, J = 7.6 Hz), 7.32–7.27 (m, 2H), 7.16 (d, 2H, J = 6.8 Hz), 6.77 (d, 2H, J = 8.1 Hz), 2.60 (t, 2H, J = 7.9 Hz), 1.61 (m, 2H), 1.36–1.29 (m, 4H), 0.91 (t, 3H, J = 6.6 Hz); ^{13}C-NMR (75 MHz, $CDCl_3$) δ 162.8 (dd, J = 253.8 Hz, J = 5.8 Hz), 162.3

(d, J = 253.4 Hz), 151.1 (ddd, J = 251.1 Hz, J = 10.4 Hz, J = 4.4 Hz), 147.4 (t, J = 9.1 Hz), 140.7 (dt, J = 256.0Hz, J = 15.5 Hz), 133.2, 127.6 (d, J = 3.2 Hz), 125.3 (d, J = 9.5 Hz), 118.6 (dd, J = 15.6 Hz, J = 6.9 Hz), 116.1 (dd, J = 15.3 Hz, J = 7.3 Hz), 111.5, 111.3 (dd, J = 17.1 Hz, J = 6.9 Hz), 98.9 (t, J = 19.9 Hz), 96.5, 92.9, 84.1, 79.9, 36.0, 31.4, 30.3, 22.5, 14.0; Anal. calcd for $C_{27}H_{18}F_6$: C, 71.05; H, 3.95. Found: C, 71.17; H, 4.11.

Characterization data for **BPEB 24**: ^{1}H-NMR (300 MHz, $CDCl_3$) δ 7.23 (d, 2H, J = 7.7 Hz), 7.14 (d, 2H, J = 7.2 Hz), 6.78 (d, 2H, J = 8.2 Hz), 2.61 (t, 2H, J = 7.5 Hz), 1.61 (m, 2H), 1.38–1.27 (m, 4H), 0.90 (t, 3H, J = 6.9 Hz); ^{13}C-NMR (75 MHz, $CDCl_3$) δ 162.9 (dd, J = 254.7 Hz, J = 6.4 Hz), 162.6 (dd, J = 254.8 Hz, J = 6.1 Hz), 155.8 (dd, J = 255.7 Hz, J = 3.6 Hz), 148.0 (t, J = 9.4 Hz), 126.3 (t, J = 15.8 Hz), 125.5 (t, J = 12.6 Hz), 123.0 (t, J = 10.6 Hz), 120.5 (q, J = 260.4 Hz), 116.0 (dd, J = 17.5 Hz, J = 6.2 Hz), 114.6 (dd, J = 18.1 Hz, J = 7.3 Hz), 111.4 (dd, J = 17.2 Hz, J = 5.5 Hz), 102.0 (t, J = 17.6 Hz), 98.4 (t, J = 19.9 Hz), 96.9, 95.5, 81.1, 79.2, 36.0, 31.3, 30.3, 22.5, 14.0; Anal. calcd for $C_{28}H_{17}F_9O$: C, 62.22; H, 3.15. Found: C, 62.18; H, 3.12.

Characterization data for **BPEB 25**: ^{1}H-NMR (300 MHz, $CDCl_3$) δ 7.19 (t, 2H, J = 6.8 Hz), 7.12 (d, 2H, J = 7.2 Hz), 6.77 (d, 2H, J = 8.2 Hz), 2.61 (t, 2H, J = 7.5 Hz), 1.61 (m, 2H), 1.34–1.32 (m, 4H), 0.90 (t, 3H, J = 6.6 Hz); ^{13}C-NMR (75 MHz, $CDCl_3$) δ 162.9 (dd, J = 254.1 Hz, J = 5.6 Hz), 162.6 (dd, J = 255.0 Hz, J = 6.4 Hz), 151.2 (ddd, J = 251.5 Hz, J = 10.6 Hz, J = 4.9 Hz), 147.9 (t, J = 9.6 Hz), 140.9 (dt, J = 256.8 Hz, J = 15.5 Hz), 125.2 (t, J = 11.6 Hz), 118.4 (td), 116.2 (ddd, J = 15.4 Hz, J = 7.6 Hz), 114.5 (dd, J = 18.5 Hz, J = 7.6 Hz), 111.3 (dd, J = 17.9 Hz, J = 5.2 Hz), 102.2 (t, J = 19.6 Hz), 98.5 (t, J = 20.1 Hz), 97.2, 95.6, 80.9, 77.8, 36.0, 31.3, 30.3, 22.5, 13.9; Anal. calcd for $C_{27}H_{17}F_7$: C, 68.35; H, 3.59. Found: C, 68.38; H, 3.76.

4. CONCLUSIONS

In summary, we have designed and synthesized BPEBs by convenient Sonogashira cross-coupling reactions, which have different numbers of side-substitute fluorine atoms on benzene rings, and alkyl chains, ethoxyl groups, fluorine atoms and trifluoromethyl groups as the end groups. The detailed investigation of the synthesized BPEBs properties have disclosed that the melting points, clearing points, nematic phase, optical anisotropy (Δn) and dielectric anisotropy (Δε) greatly depend on both the numbers of side-substitute fluorine atoms and structures of the end groups. On the basis of the obtained properties of the synthesized BPEBs, some of them have been expected to have high potential application as the compositions in blue phase liquid crystals. Therefore, a mixture of blue phase liquid crystal has been prepared with a relative wide blue phase temperature range of 8 °C. The obtained results have implied that the synthesized BPEBs will certainly be important in the development of new types and properties of LCs. Further study on the application of BPEBs in making other new types of LCs is underway in our laboratory.

ACKNOWLEDGMENTS

Financial support from the National High Technology Research and Development Program, China (No. 2011AA02A209) is gratefully acknowledged.

CONFLICTS OF INTEREST

The authors declare no conflict of interest.

REFERENCES

1. Carlton, R.J.; Hunter, J.T.; Miller, D.S.; Abbasi, R.; Mushenheim, P.C.; Tan, L.N.; Nicholas, L.; Abbott, N.L. Chemical and biological sensing using liquid crystals. Liq. Cryst. Rev **2013**, 1, 29–51.

2. Tadwee, I.; Shahi, S.; Ramteke, V.; Syed, I. Liquid crystals pharmaceutical application: A review. Int. J. Pharm. Res. Allied Sci **2012**, 1, 6–11.

3. Goodby, J.W. The nanoscale engineering of nematic liquid crystals for displays. Liq. Cryst **2011**, 38, 1363–1387.

4. Kitzerow, H.-S. Blue phases come of age: A review. Proc. SPIE **2009**.

5. Hong, H. Analysis of focal length of blue-phase liquid crystal (BPLC) cylindrical lens for the light of the various incident angles and polarizations. Liq. Cryst **2013**, 40, 450–457.

6. Yamamoto, S.-I.; Haseba, Y.; Higuchi, H.; Okumura, Y.; Kikuchi, H. Lattice plane control of liquid crystal blue phase. Liq. Cryst **2013**, 40, 639–645.

7. Li, P.; Sun, Y.; Wang, Q. A transflective and viewing angle controllable bluephase liquid crystal display. Liq. Cryst **2013**, 40, 1024–1027.

8. Tour, J.M. Molecular electronics. Synthesis and testing of components. Acc. Chem. Res **2000**, 33, 791–804.

9. Bunz, U.H.F. Poly (aryleneethynylene)s: Syntheses, properties, structures, and applications. Chem. Rev **2000**, 100, 1605–1644.

10. Levitus, M.; Schmieder, K.; Ricks, H.; Shimizu, K.D.; Bunz, U.H.F.; Garcia-Garibay, M.A. Steps to demarcate the effects of chromophore aggregation and planarization in poly(phenyleneethynylene)s. 1.

Rotationally interrupted conjugation in the excited states of 1,4-bis(phenylethynyl)benzene. J. Am. Chem. Soc **2001**, 123, 4259–4265.

11. Beeby, A.; Findlay, K.; Low, P.J.; Marder, T.B. A re-evaluation of the photophysical properties of 1,4-bis(phenylethynyl)benzene: A model for poly(phenyleneethynylene). J. Am. Chem. Soc **2002**, 124, 8280–8284.

12. Schwab, P.F.H.; Smith, J.R.; Michl, J. Synthesis and properties of molecular rods. 2. Zig-zag rods. Chem. Rev **2005**, 105, 1197–1279.

13. Wu, S.-T.; Hsu, C.-S.; Shyu, K.-F. High birefringence and wide nematic range bis-tolane liquid crystals. Appl. Phys. Lett **1999**, 74, 344–346.

14. Tanaka, T.; Sekine, C.; Ashida, T.; Ishitobi, M.; Konya, N.; Minai, M.; Fujisawa, K. Highly anisotropic molecular materials for LCD. Mol. Cryst. Liq. Cryst **2000**, 346, 209–216.

15. Long, T.M.; Swager, T.M. Triptycene-containing bis(phenylethynyl) benzene nematic liquid crystals. J. Mater. Chem **2002**, 12, 3407–3412.

16. Liao, Y.-M.; Chen, H.-L.; Hsu, C.-S.; Gauza, S.; Wu, S.-T. Synthesis and mesomorphic properties of super high birefringence isothiocyanato bistolane liquid crystals. Liq. Cryst **2007**, 34, 507–517.

17. Lydon, D.P.; Albesa-Jové, D.; Shearman, G.C.; Seddon, J.M.; Howard, J.A.K.; Marder, T.B.; Low, P.J. The synthesis and liquid crystalline behaviour of alkoxy-substituted derivatives of 1,4-bis(phenylethynyl)benzene. Liq. Cryst **2008**, 35, 119–132.

18. Dziaduszek, J.; Kula, P.; Dąbrowski, R.; Drzewiński, W.; Garbat, K.; Urban, S.; Gauza, S. General synthesis method of alkyl-alkoxy multifluorotolanes for negative high birefringence nematic mixtures. Liq. Cryst **2012**, 39, 239–247.

19. Miao, Z.-C.; Wang, D.; Zhang, Y.-M.; Jin, Z.-K.; Liu, F.; Wang, F.; Yang, H. Asymmetrical phenyldiacetylenes liquid crystalline compounds with high birefringence and characteristics of selective reflection. Liq. Cryst **2012**, 39, 1291–1296.

20. Zhang, Y.-M.; Wang, D.; Miao, Z.-C.; Jin, S.-K.; Yang, H. Novel high birefringence bistolane liquid crystals with lateral fluorosubstituent. Liq. Cryst **2012**, 39, 1330–1339.

21. Herman, J.; Dziaduszek, J.; Dąbrowski, R.; Kędzierski, J.; Kowiorski, K.; Dasari, V.S.; Dhara, S.; Kula, P. Novel high birefringent isothiocyanates based on quaterphenyl and phenylethynyltolane molecular cores. Liq. Cryst **2013**, 40, 1174–1182.

22. Yatabe, T.; Okumoto, H.; Kawanishi, Y.; Inoue, T. Charge-carrier transport in 1,4-bis(phenylethynyl)benzene derivatives exhibiting crystal mesophases. Chem. Lett **2013**, 42, 764–766.

23. Imrie, C.T.; Taylor, L. The preparation and properties of low molar mass liquid crystals possessing lateral alkyl chains. Liq. Cryst **1989**, 6, 1–10.

24. Iwata, T.; Suzuki, K.; Higuchi, H.; Kikuchi, H. A method for enlarging the Kerr constant of polymer-stabilised blue phases. Liq. Cryst **2009**, 36, 947–951.

CHAPTER 10

Gas Safety for TFT-LCD Manufacturing

Eugene Y. Ngai[1] and Jenq-Renn Chen[2]

1 Chemically Speaking LLC, New Jersey, USA

2 Department of Safety, Health and Environmental Engineering, National Kaohsiung First University of Science & Technology, Kaohsiung, Taiwan

1. INTRODUCTION

The fabrication of TFT-LCD panel includes the growing and etching of a-silicon and silicon nitride films. These fabrication processes utilize significant amounts of silane, phosphine, ammonia, chlorine, boron trichloride, nitrogen trifluoride, fluorine, and hydrogen which are highly flammable, reactive, corrosive, and/or toxic. The amount used and scale of the supply system for these gases are far larger than for other tech industries such as the semiconductor and photovoltaic industries. Accidental leaks and fires of these gases are the major safety concern in the TFT-LCD fabs.

This paper first reviews the hazardous properties of the gases used in the TFT-LCD manufacturing processes. The best practices for handling these hazardous gases are then described. Finally, the past incidents and emergency response actions are also reviewed.

2. HAZARDOUS PROPERTIES OF SPECIALTY GASES

The TFT-LCD manufacturing processes utilize significant amount of gases for thin-film deposition and etching. These include silane, phosphine, ammonia, and hydrogen for polycrystalline silicon and silicon nitride thin film deposition in a plasma-enhanced chemical vapor deposition (PECVD) reaction chamber. In addition, nitrogen trifluoride or fluorine is used in the PECVD chamber cleaning. Chlorine and sulfur hexafluoride are used in the dry etching of thin film. The hazards of these gases can be classified as pyrophoric, flammable and oxidizing gases. There are other bulk gases such as nitrogen, argon and helium which are used for inerting or purging and will not be discussed here owing to their low risks.

2.1 Silane

Silane or silicon tetrahydride (SiH4) is the most common silicon source used in TFT-LCD manufacturing. It is a highly flammable gas and has a very wide flammable range, from 1.37% to 96%. Silane is colorless and odorless although it has been reported to have a prudent odor and a reported time weighted threshold limit value (TLV-TWA) of 5 ppm by ACGIH. However, this data is based on analogy of another hydride gas Germane rather than the actual toxicological data. In fact, silane has no odor and its toxicity is low with a high median lethal concentration (LC50) of 9,600 ppm for rat at 4 hours exposure. The prudent odor comes from the impurity of trichlorsilane during early silane production. Silane is used as a compressed gas. However, silane has a critical temperature of -3.4°C. It is possible that liquefaction may occur in cold storage or during expansion cooling from compressed sources. A silane supply system must be designed properly to prevent unexpected pressure surge from liquefaction and vaporization.

Silane is also a pyrophoric gas that normally ignites upon contact with air. Silane has a reported autoignition temperature of -50~-100°C. Autoignition of silane in air has also been reported down to -162°C, depending on the oxygen concentration of the mixture (Baratov et al., 1969). However, the autoignition temperature only denotes the temperature above which a given fuel-air mixture will autoignite in a closed vessel. It however does not refer to details of the autoignition process and how autoignition is affected by flow. The major potential hazard of silane is however not in its pyrophoricity but rather in its unpredictable ignition behavior in accidental releases. A silane release from a pressure source has been known not to lead to spontaneously ignition (Koda, 1992) and frequently delayed ignition occurs when the release is shut-off resulting in a "pop". In a semi-confined space with gas accumulation, the pop can lead to a gas explosion with significant damage (Ngai et al., 2007).

The mechanism of delayed ignition has been studied by Tsai et al. (2011) which demonstrated that silane release without prompt ignition was most likely caused by quench of the reactive kernel from the flow strain or scalar dissipation accompanied by the large velocity and concentration difference between the release gas and the ambient air. With diminishing release velocity, the flow strain reduces along with quench of the reactive kernel and ignition then occurs at a critical exit velocity. The ignition at reducing velocity may ignite the released gas and create a significant combustion or explosion. The critical exit elocity of indefinitely delayed ignition lies between 0.3 m/s to 4.3 m/s for vent diameter of 2.03 ~4.32 mm as shown in Figure 1. These velocities are at least two orders of magnitude smaller than the velocity from a pressurized source which in most cases reaches sonic velocity. Furthermore, Figure 1 also highlights that the critical velocity is also decreases with decreasing vent size.

There are factors other than vent size affecting the critical exit velocity. Among them, the temperature, moisture and silane combustion powder are the most noticeable factors. At higher temperature, the reactive kernel is stronger and thus the critical exit velocity is larger (Liang et al., 2010). Removing the moisture in the air also results in significantly higher critical exit velocity indicating an inhibitory role of moisture on silane ignition (Liang et al., 2010). The inhibitory action of moisture on the silane autoigniton in air is also in consistent with other fuel such as hydrogen and methane. Silane combustion produces white to brown powders as shown in Figure 2. These powders are also known to promote ignition but its exact role remains to be studied.

The presence of a critical exit velocity for prompt ignition of silane release has important implications in the safety of silane operations. First of all, almost all current silane uses in the semiconductor, TFT-LCD, and photovoltaic industries are supplied in the form of pressurized cylinders with pressure up 12.5 MPa. The high pressure silane is then regulated to 0.8 MPa for the supply tube into the cleanroom and then further regulated down to 0.3~0.45 MPa before feeding into the process chamber. Thus, almost all potential leak points in silane utilization have pressure and possible leak velocity higher than the reported critical velocity for prompt ignition. In addition, the potential leak points are the cylinder valve, tubing, and tube connections (Huang and Ngai, 2006). The potential release size will be much smaller than 2.03 mm, except for a catastrophic full bore tube rupture. Thus, delayed ignition should be considered as the usual case rather than a rare case in most silane operations.

2.2 Phosphine

Phosphine (PH3) is a highly toxic and pyrophoric liquefied compressed gas. Phosphine is used primarily as a phosphorus dopant (N-type dopant) in the TFT LCD industry. It is typically supplied as a mixture with hydrogen.

Pure phosphine is a colorless, pyrophoric liquefied gas. Although pure samples of phosphine are odorless, the gas often has an odor of garlic or decaying fish due to the presence of substituted phosphine and diphosphine (P2H4). It is a highly toxic gas with a PEL of 300 ppb, an LC50 of 20 ppm for rat at 1 hour exposure.

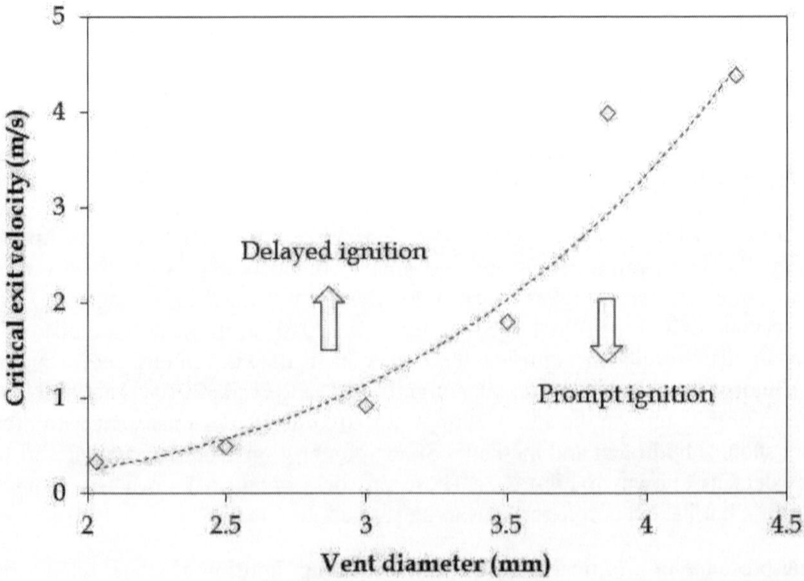

Figure 1. Critical exit velocity for delayed/prompt ignition as a function of vent diameter.

Phosphine has a low auto-ignition temperature <0°C so it may ignite spontaneously on contact with air. Autoignition of phosphine released into air is also subjected to effects by temperature, moisture, and flow strain just like silane. For example, autoignition of phosphine is reported to occur only in dry air but not in moist air (Brittion, 1990). However, there is still a lack of detailed studies on the exact conditions for these effects. It also has an extremely wide flammable range of 1.6% – 95% in air. It burns in air with an orange flame forming white phosphorous pentoxide as a byproduct as shown in Figure 3(a), which is a severe respiratory irritant. Under certain conditions, phosphine will form red phosphorus as a byproduct in a fire as shown in Figure 3(b).

Acute exposures to phosphine cause respiratory tract irritation that attacks primarily the cardiovascular and respiratory systems causing peripheral vascular collapse, cardiac arrest and failure, and pulmonary edema. Inhalation is the primary exposure route, and there is no known dermal exposure route. Symptoms may include restlessness, irritability, drowsiness, tremors, vertigo, double vision (diplopia), ataxia, cough, dyspnea, retrosternal discomfort, abdominal pain, and vomiting. There is no antidote for phosphine exposures; treatment is supportive. Guidance on proper medical treatment is available from the US Health Department Agency for Toxic Substances and Disease Registry (ATSDR) sheet on phosphine.

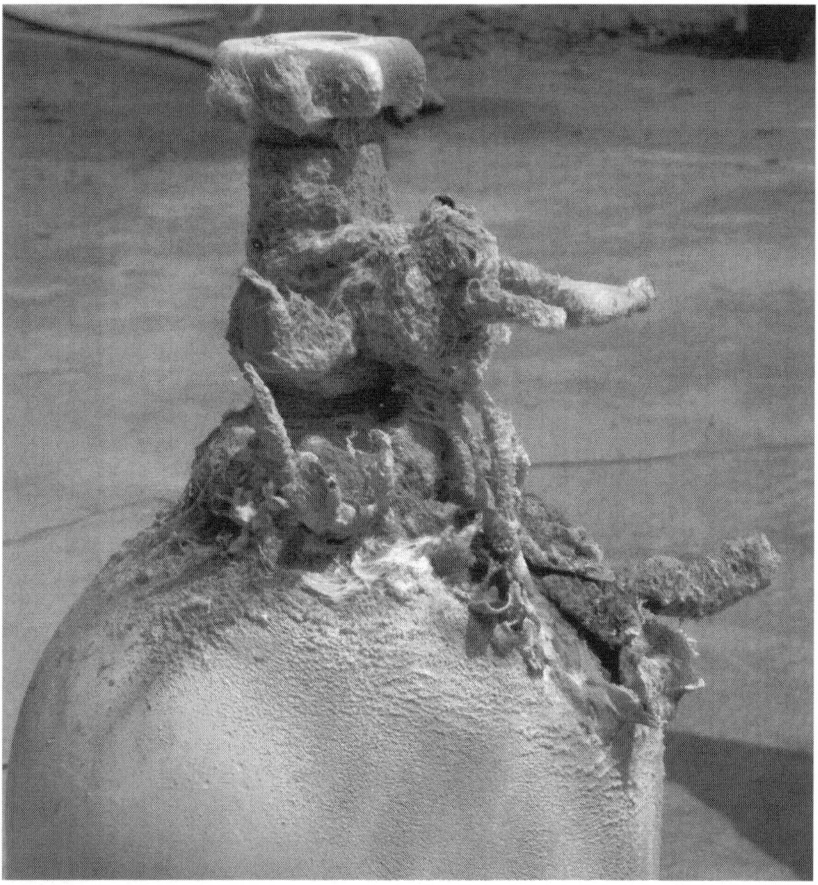

Figure 2. Typical silane combustion powder.

Transportation regulations prohibit the supply of pure phosphine in cylinders larger than 50 liters since it is highly toxic. As a dopant, phosphine is supplied to the process as a low concentration mixture in order to precisely control the diffusion rate for the TFT-LCD processes. As a mixture, the gas is no longer classified as highly toxic, therefore it is supplied in bulk high cylinders of 450 liters or larger.

Some gas suppliers are also supplying gas mixing systems which will dynamically blend the phosphine from a pure source cylinder with hydrogen. A single cylinder will supply 200 cylinders of 0.5% concentration.

Figure 3. Typical phosphine combustion flame and powder. (a) Orange flame and white phosphorous pentoxide powder (b) Red phosphorus powder.

2.3 Hydrogen

Hydrogen is a colorless, odorless, tasteless, highly flammable gas. It is also the lightest weight gas. It is much less dense than air and can disperse rapidly or accumulate in the upper sections of enclosed spaces

Hydrogen has a wide flammability range, 4% to 75% in air, and the small amount of energy required for ignition necessitate special handling to prevent the inadvertent mixing of hydrogen with air. Sources of ignition such as sparks from electrical equipment, static electricity, open flames, or extremely hot objects should be eliminated. Hydrogen and air mixtures within the flammable range can explode and may burn with an almost invisible flame. Although hydrogen is not pyrophoric like silane or phosphine, hydrogen release from a tube or a crack under high pressure may lead to spontaneous ignition. This is attributed to the low ignition energy, high blow-off velocity and the adiabatic compression from the shock wave of the release (Dryer et al., 2007)

Gaseous hydrogen may be supplied in cylinders or in tubes that are designed and manufactured according to applicable codes and specifications for the pressures and temperatures involved. The pressure rating and internal volume of a container determines the quantity of hydrogen it can hold. Cylinders may be used individually or can be manifolded together to allow for a larger gas storage volume. Tubes are mounted on truck- trailer chassis or in ISO frames for transportation and are referred to as tube trailers or tube modules, respectively.

2.4 Ammonia

Ammonia (NH3) is a toxic and corrosive liquefied compressed gas. It is used to grow a silicon nitride layer in the TFT-LCD industry. Pure ammonia is a white liquefied gas. It is a strongly irritating gas. It is a toxic gas with a PEL of 50 ppm and an LC50 of 7,338 ppm. Ammonia is lighter than air (specific gravity: 0.59) and is highly soluble in water. Ammonia is thermally stable.

Ammonia has a very narrow flammable range of 16% to 25% and minimum ignition energy (MIE) 40,000 times (680 mJ) higher than that of hydrogen (0.017 mj). The common flammable gases such as propane have a MIE of 0.25 mJ which is still many times less than amonnia. The high lower flammable limit (LFL) and lighter than air vapor density makes it unlikely for an ammonia release to reach concentrations high enough to burn in open areas. It is thus classified as non-flammable for transportation purposes. The national building and fire regulations however classify it as flammable in storage and use.

As a flammable gas, ammonia is reactive with oxidizer gases such as nitrous oxide or oxygen. Since ammonia has an LC50 higher than 5,000 ppm (threshold for toxic gas), the United States classifies it as a non-toxic non-flammable gas. All other countries classify ammonia as a toxic gas. Ammonia is a severe irritant with a sharp pungent odor that will have an immediate effect on moist tissues, eyes, upper respiratory system and skin. Guidance on proper medical treatment is available from the US Health Department Agency for Toxic Substances and Disease Registry (ATSDR) sheet on Ammonia.

Ammonia is very corrosive to low alloy carbon steel, causing stress corrosion. Ammonia is also corrosive to viton, zinc and copper. Bulk ammonia is supplied in low or high pressure bulk containers (250 liter - 950 liter). Some facilities use 40,000 lb ISO Modules that are sited like the nitrogen trifluoride or silane modules for continuous supply. It is vaporized for use on demand.

2.5 Nitrogen trifluoride

Nitrogen trifluoride (NF3) is a colorless, odorless, nonflammable, oxidizing compressed gas. Nitrogen trifluoride is the most common reactor cleaning gases used in the TFT-LCD industry because of its advantages, such as high etching rates, high selectivity, carbon-free etching, and minimal residual contamination. It is first decomposed in a remote plasma chamber into a reactive fluorine before fed to the reaction chamber where it is used to react solid deposits on the reactor walls to form gaseous byproducts such as silicon tetrafluoride.

Nitrogen trifluoride is not toxic, it has an LC50 of 6,700 ppm (1 hr. rat). It is not hazardous by skin contact and is a relatively minor irritant to the eyes and mucous membranes. Exposure of rats to a 100 ppm concentration for 7 hrs per day, 5 days per week over a 19 week period caused minor pathological changes to the liver and kidney. No other effects were noted. The ACGIH established the current TLV-TWA of 10 ppm based on 1/10 of this value. Some countries classify nitrogen trifluoride as a toxic gas because it has a TLV- TWA <200 ppm.

Under ambient temperatures and pressures, nitrogen trifluoride is inert. However it can become an extremely reactive under certain conditions. It is a strong oxidizer that is violently explosive at elevated pressures with flammable gases such as hydrogen or silane. Purification of nitrogen trifluoride using a solid absorbent media such as molecular sieve is to be avoided. Fires and explosions have occurred.

Nitrogen trifluoride is reported to be a significant greenhouse warming gas with a Greenhouse Warming Potential (GWP) >10,000 as compared to carbon dioxide. Properly design remote plasma cleaning system will react 100% of the nitrogen trifluoride, so it has a GWP comparable to fluorine under these conditions.

2.6 Chlorine

Chlorine (Cl2) is a toxic and corrosive liquefied compressed gas. It is also an oxidizing gas. Chlorine is used primarily as an etching gas in the TFT-LCD industry. Pure chlorine is a yellow-green liquefied gas. It is a strongly irritating and highly toxic gas with a PEL of 0.5 ppm and an LC50 of 293 ppm. Chlorine is much heavier than air (specific gravity: 2.49) and is slightly soluble in water. Chlorine is thermally stable.

Chlorine is a severe irritant with a pungent odor that will have an immediate effect on moist tissues, eyes, upper respiratory system and skin. Chronic exposure will corrode teeth, in As a flammable gas, ammonia is reactive with oxidizer gases such as nitrous oxide or oxygen. Since ammonia has an LC50 higher than 5,000 ppm (threshold for toxic gas), the United States classifies it as a non-toxic non-flammable gas. All other countries classify ammonia as a toxic gas. Ammonia is a severe irritant with a sharp pungent odor that will have an immediate effect on moist tissues, eyes, upper respiratory system and skin. Guidance on proper medical treatment is available from the US Health Department Agency for Toxic Substances and Disease Registry (ATSDR) sheet on Ammonia.

Ammonia is very corrosive to low alloy carbon steel, causing stress corrosion. Ammonia is also corrosive to viton, zinc and copper. Bulk ammonia is supplied in low or high pressure bulk containers (250 liter - 950 liter). Some facilities use 40,000 lb ISO Modules that are sited like the nitrogen trifluoride or silane modules for continuous supply. It is vaporized for use on demand.

2.7 Fluorine

Fluorine is a highly toxic gas with extreme oxidation and corrosion potentials. It has a sharp, pungent odor that can be detected by most people at very low levels (0.1 ppm). It has a TLV-TWA of 1 ppm and an LC50 of 185 ppm (1 hr. rat). Chronic fluorine exposures at low levels may cause fluorosis or abnormal calcium accumulation in bone structure. Fluorine is highly reactive with moisture, forming hydrofluoric acid which is extremely corrosive to human tissue. Fluorine will have an immediate effect on the eyes and respiratory system.

Medical treatment of exposure is specific and specialized. Releases into air can form varying concentrations of hydrofluoric acid, depending on conditions. When the aqueous hydrofluoric acid concentration is below 50%, it has delayed (up to six hours) symptoms of exposure. In a fire, fluorine will react to form other toxic fluoride compounds.

Fluorine is the most reactive of all elements and the most powerfully know oxidizing agent. Fluorine is able to react with almost all elements and compounds depending on pressure and temperature, with the exception of lighter noble gases (e.g. helium, argon),inorganic fluorides of the highest valency, and perfluorinated organic compounds. It reacts with many organic substances even at room temperatures and often accompanied by combustion and possible explosion (European Industrial Gases Association, 2011). Therefore special steps, such as special cleaning and passivation procedures, must be taken to protect systems.

Fluorine reacts vigorously at ambient temperatures with most metals. The reaction intensity depends upon the surface area of the metal. Powdered metals or fine wires may react violently. Most inorganic materials react with fluorine; water forms hydrofluoric acid, and salts convert to fluorides. Contact with organic materials generally results in ignition or violent explosion.

Due to its extreme reactivity, national transportation regulations typically limit the total amount of fluorine in a 50 liter cylinder to 3 kg or to a pressure of 30 bar. Pure fluorine cylinders can be fairly reactive at 30 bar. Some users use a mixture which is less reactive. Testing by Air Products and Chemical Inc. (2004) has shown that fluorine is 2.5 times more reactive than oxygen, when it is mixed in a 20% concentration with nitrogen it will not ignite a 1⁄4" carbon steel rod at 13,790 kPa.

Testing by Ngai (2005) has shown that pure fluorine even at a pressure as low as 1 atm (101 kPa) can cause 316 stainless steel tubing to burn. Pure fluorine did not react immediately with the raw chicken. To better simulate human skin, some human hair was placed on the chicken skin. This immediately ignited and started the skin on fire as shown in Figure 4. The reaction was so hot that it ignited the stainless steel tubing, which sprayed the chicken with molten stainless steel. This continued to react until the fluorine flow was stopped (Ngai, 2005).

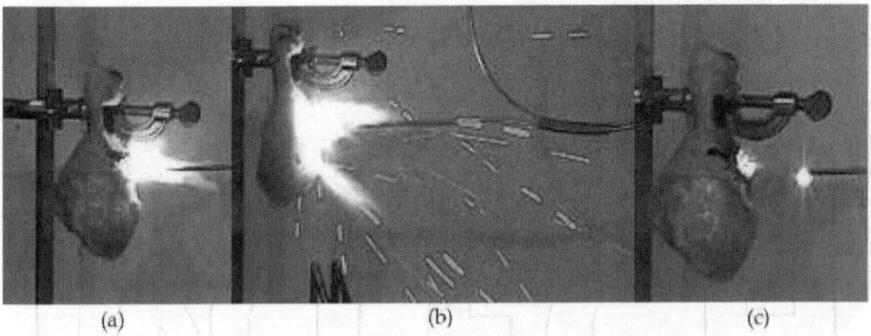

Figure 4. Reaction of fluorine and chicken skin. (a) The hair on the skin reacts with the fluorine and ignites the chicken skin. (b) Skin burns hot enough to ignite the stainless steel tubing, which is sprayed at the chicken. (c) The tubing continues to react until the flow is stopped.

Despite its toxicity and reactivity, fluorine is being increasingly used as a reactor cleaning gas owing to less energy requirement, lower Greenhouse Warming Potential, and faster clean time (Stockman, 2009). To supply pure fluorine, small generators are being installed and managed by gas suppliers. These limit the amount of fluorine stored. A typical fluorine generator is an electrolytic cell which dialysis hydrogen fluoride to form fluorine and hydrogen (Stockman, 2009).

3. BEST PRACTICE FOR HANDLING

3.1 Silane

The first step in safe handling of silane is to comply with relevant regulatory & industrial standards such as Compressed Gas Association (CGA), Semiconductor Equipment and Materials International (SEMI), National Fire Protection Association (NFPA), Factory Mutual (FM), European Industrial Gas Association (EIGA), Asian Industrial Gas Association (AIGA), Transportable Pressure Equipment Directive (TPED), Transportation of dangerous goods (ADR), SEVESO, ATEX, etc. The installation and operation practices are however not necessarily detailed in the above codes or standards. Rufin (2011) has presented a comprehensive list of best practices for silane supply system. The key features are summarized below.

From the perspectives of preventing silane explosion, the most effective measure is preventing the accumulation of unignited silane through the use of forced ventilation or open air storage. It is also important to detect an unignited release by installing a gas sensor. Flame (UV/IR) detector is also needed for detecting leaks with autoignition. Sprinklers are needed for cooling the cylinders and system. Operators must be also be protected by barriers and flame resistant protection suit.

Other minor but equally important issues include:

- Welding for all indoor distribution systems
- Joule-Thomson compensation for high flow >50 L/min
- Automation and remote control
- Prevention of "domino effect" by steel plate or firewall between sources
- Normally closed pneumatic shut off valves
- Dedicated purge sources with automated purge cycles
- High pressure test and He outboard leak test
- Emergency remote shut down

For high flow rate greater than 50 L/min, large pressure drop will produce liquid silane mixture causing failure and leaks mostly in regulators. Thus, use of 2 regulators in series and heaters to compensate Joule-Thomson expansion cooling and prevent condensation.

3.2 Fluorine

As fluorine reacts with almost everything, corrosion of materials in contact with fluorine is dependent upon the fluoride formed with the material. If the fluoride is volatile or in the form of loose flakes, new material will constantly be exposed to the fluorine and will be continuously eroded until it burns through the material. For this reason material selection is extremely important. Materials that form a good fluoride coating will be protected from further attack.

Systems for fluorine service must not only be fabricated of the proper material, the design of the system must consider the following factors:

- Adiabatic compression - Adiabatic compression and heating is caused by the rapid pressurization of a system where the gas absorbs the energy and the gas temperature rises. This heating occurs at the point of compression or the point where the flow of gas is stopped, such as at a valve or regulator seat. Depending on the material in use where the hot gas impinges, the heat may be sufficient to ignite the material.
- High flow – High flow velocity increases the potential for a particle to impinge onto a surface with sufficient force to ignite it. A common area where this occurs is at a direction of flow change such as an elbow.

- Reactive materials - If the valve or regulator has a seat made of a nonmetallic material, it may be more prone to ignition because nonmetals typically have lower ignition temperatures than metals.
- Drying – removal of moisture from the system surface is essential
- Cleaning - to remove contaminants such as oil and particles from the system. These can be ignited easily generating sufficient heat to propagate a reaction to the system.
- Passivation – Fluorine passivation of metals is a pretreatment of the components in order to stabilize the wetted surfaces of the system by formation of a self-limiting protective fluoride film on the surface of the metal. The typical film thickness range from thickness of 5 Å (0.0005 microns) to 50 Å (0.005 microns) depending on the metal (Air Force Rocket Propulsion Laboratory, 1967).
- Friction - from a component that is malfunctioning or operating poorly can generate heat. Friction between two materials generates fine particles, which may ignite from the heat generated.

3.3 Other specialty gases

Silane and fluorine represent the two extremes, the flammable and the oxidizer, of the specialty gases used in TFT-LCD manufacturing. All other gases bear similar but less extensive practices and protection measures as these two gases. For example, gas detectors are needed for all gases except for nitrogen trifluoride. Flame (UV/IR) detectors and sprinklers are needed for all flammable gases. High pressure test and He outboard leak test with emergency remote shut down are also typical practices for all gases.

4. INCIDENTS AND EMERGENCY RESPONSE

4.1 Silane

There are numerous silane leaks in the forty years of silane use (Huang and Ngai, 2006) mostly resulted from poor procedures, practices, and equipment. The reactions from the leaks were reported to be a fire, minor "pops" to explosions. The "pops" can occur when a small amount of Silane trapped behind the valve cap or pigtail is released rapidly without prompt ignition. In a few cases they have been severe enough to cause eardrum rupture if the amount of accumulated silane is significant. Thus, the past silane leaks do support the fact that a leak with indefinitely delayed ignition is commonly observed in silane operations.

Although the release with indefinitely delayed ignition were mostly small and had finite impact, the unignited release may turn into large release with significant impact. The following major incidents were known where silane was released without immediate ignition and exploded after a delay:

- 1976, Germany, 1 fatality
- 1989, Japan, 1 fatality & 1 injury
- 1990, Japan, 1 fatality & 3 injuries
- 1992, US, 1 injury
- 1996, Japan, 1 fatality
- 2003, US, no injuries
- 2005, Taiwan, 1 fatality
- 2007, India, 1 fatality

Among these incidents, the incident in Taiwan in 2005 has been described in detail by Chen et al. (2006), Chang et al. (2007) and Peng et al. (2008). The cause of leak is a failed silane cylinder valve with loosened valve stem retainer. Silane leak from such a loosened retainer has been shown to have delayed ignition by Ngai et al. (2007). The operator in this case did not notice the unignited silane release from the loosened retainer and continued to turn the valve handwheel until eventually the retainer, valve stem, and valve diaphragm were all detached. The rapid silane release with ignition at end of release resulted in a fatal vapor explosion.

Emergency response to a silane incident is the most challenging one among all gas incidents (Ngai, 2011). The greatest danger is a leak and no fire. As ignition is inevitable at the end of leak, a pop or explosion should be expected. Use surveillance camera to make initial assessment if possible. Remotely review the gas detector response over time on a graph to insure that it is not sensor drift or electronic spike. Evacuation should be done first if there is a gas detector alarm but without visual confirmation of flame. The responders should wear full personal protection equipment, including fire proof suit, glove, hardhat, ear plug etc. and approach the leak with great care. The leak could ignite as the pressure decays in the source cylinder or an air flow disturbance is created. Assess the leak rate with a portable gas detector through the gas cabinet duct. For a small leak, it might be safe to remotely shutoff the supply but expect ignition and a pop at shutoff. Slow shutoff of the valve will minimize the amount of silane when it ignites. A leak with a gas detector reading below the maximum scale, normally about 50 ppm, may be considered as a small leak. For a larger leak, activate sprinkler or apply water spray from a distance to the leak point to disperse the gas cloud and reduce the extent of potential explosion. Always prepare for possible explosion.

If the leak is burning, cool the adjacent area with a water spray. Do not spray water directly onto the fire as this could put it out and create an explosive environment. The heat could melt the fusible metal in the pressure relief device (PRD). The PRD will not activate unless the internal pressure reaches 27.6 MPa (4000 psig). Water spray from the deluge system onto the tubes will prevent this from ever reaching this temperature. A leak on fire can plug from the solids formed. Solids can also trap silane which can still be reactive. Even after a fire and the cylinder has vented to zero gauge pressure, there is still one atmosphere of gas in the cylinder. Emergency response teams typically will dilute the remaining silane by pressurizing the leaking cylinder with nitrogen to 0.79 MPa (100 psig) and venting 3 times. This will dilute the silane concentration to 2000 ppm. At this concentration and pressure, the cylinder can be shipped back as a non-hazardous package.

4.2 Hydrogen

The authors are aware of six hydrogen fires with one of the fires resulting in rupture of the hydrogen tube. The fire resulted from the detachment of an aging copper tube to the cylinder pack in a hydrogen filling station. The hydrogen ignited during tube detachment from a filling manifold and resulted in jet fire which impinged on a nearby tube trailer. The tube trailer was parked without protection of sprinkler system and two tubes were ruptured in less than 5 minutes of flame impingement. The most important emergency response after the tube explosion was continuous cooling and venting of hydrogen in the remaining tubes.

4.3 Nitrogen trifluoride

There have been reported incidents of nitrogen trifluoride with flammable oils used in vacuum pumps. Systems for nitrogen trifluoride must be designed similar to the requirements for fluorine in 3.2. Adiabatic compression and heating can initiate the decomposition reaction of nitrogen trifluoride into reactive fluorine. Incidents have occurred with improperly designed or cleaned systems during routine activities such as pressurizing the system. Internationally the maximum fill limits for nitrogen trifluoride cylinders is 0.5 kg/liter. At 21°C the cylinder pressure will be 10.78 MPa. Quickly opening the cylinder valve into a system with a small dead volume will cause adiabatic compression and heating to 307°C. In the presence of a reactive nonmetal component or a contaminant, this can initiate the decomposition reaction. Given sufficient energy and pressure, the reaction can cause the metal system to also become involved. Figure 5 shows a burnt valve in a nitrogen trifluoride line caused by adiabatic compression.

4.4 Chlorine

Chlorine is very corrosive to carbon steel and stainless steel. With the moisture in the air it forms hypochlorous (HClO) and hydrochloric (HCl) acids which will corrode the steel as shown in Figure 6. A corroded connection generally leads to leak. A good design suply system should pick up the leak by detector and shutdown the system. If the system is not designed properly, the chlorine may continuous to leak and affect the cleanroom. The author is aware of a chlorine leak from a tube connection in an IC fab that was not detected until one hour later. The chlorine cloud speaded and significantly affected the cleanroom and production tools.

Figure 5. A burnt valve in a nitrogen trifluoride line caused by adiabatic compression.

4.5 Other specialty gases

The authors are not aware of any phosphine, ammonia, or fluorine fatalities or fires related to semiconductor or TFT-LCD uses.

Figure 6. Green hypochlorous and hydrochloric acids from chlorine corrosion.

5. CONCLUSIONS

The TFT-LCD manufacturing processes utilize significant amount of gases ranging from highly flammable to strong oxidizer gases. Special cares must be taken in handling these specialty gases to prevent fire, explosion and toxic release.

This paper has reviewed the hazardous properties and potential incidents of these specialty gases. The best practices for handling these gases are then described. The emergency response actions are also reviewed. It is sincerely hope that the current summary may bring an integrated view on the complicated and delicate specialty gas supply system, and preventing future incidents from occurring.

REFERENCES

1. Air Force Rocket Propulsion Laboratory (1967). Halogen Passivation Procedural Guide, Technical Report AFRPL-TR-67-309, December 1967.

2. Air Products and Chemical Inc. Safetygram #25: 20% Fluorine in Nitrogen, June 2004.

3. Baratov, A. N.; Vogman, L. P. & Petrova, L. D. (1969). Explosivity of monosilane-air mixtures. *Combustion, Explosion and Shock Waves*, Vol. 5, No. 4, (September 1969), pp. 592–594, ISSN 0010-5082.

4. Britton, L. G. (1990). Combustion hazards of silane and its chlorides, *Plant/Operations Progress*, Vol. 9, No. 1, (January 1990), pp. 16-38, ISSN 0278- 4513.

5. Chang, Y. Y., Peng, D. J., Wu, H. C., Tsaur, C. C., Shen, C. C., Tsai, H. Y., & Chen, J. R. (2007). Revisiting of a silane explosion in a photovoltaic fabrication plant, *Process Safety Progress*, Vol. 26, No. 2, (June 2007), pp. 155-157, ISSN 1066-8527.

6. Chen, J. R. , Tsai, H. Y., Chen, S. K., Pan, H. R., Hu, S. C., Shen, C. C., Kuan, C. M., Lee, Y. C., & Wu, C. C. (2006). Analysis of a silane explosion in a photovoltaic fabrication plant, *Process Safety Progress*, Vol. 25, No. 3, (September 2006), pp. 237-244, ISSN 1066-8527.

7. Dryer, F. L., Chaos, M., Zhao, Z., Stein, J. N., Alpert, J. Y. & Homer, C. J. (2007). Spontaneous ignition of pressurized release of hydrogen and natural gas into air. *Combustion Science and Technology*, Vol. 179, No. 4, (April 2007), pp. 663–94, ISSN 0010-2202.

8. European Industrial Gases Association (2011). Code of Practice: Compressed Fluorine and Mixtures with Inert Gases, European Industrial Gas Association, Globally Harmonized Document, IGC 140/11, 2011

9. Huang, K. P. P. & Ngai, E. Y. (2006). Silane Safety and Silane Incidents, *Silane Safety and ER Seminar*, Kaohsiung, Taiwan, May 17-19.

10. Koda, S. (1992). Kinetic aspects of oxidation and combustion of silane and related compounds, *Progress in Energy Combustion Science*, Vol. 18, No. 6, (November 1992), pp. 513-528, ISSN 0360-1285.

11. Liang, R. L., Wu, S. Y., Tseng, N. J., Ku, C. W., Tsai, H. Y. & Chen, J. R. (2010). Effects of Temperature and Moisture on the Ignition Behavior of Silane Release into Air, *8th International Symposium on Hazards, Prevention, and Mitigation of Industrial Explosions*, Yokohama, Japan, September 6-10 , 2010.

12. Ngai, E. Y. (2005). Release Testing of Reactive Fluoride Reactor Cleaning Gases, *Semicon China*, Shanghai, China, March 14 & 15, 2005.

13. Ngai, E. Y. (2011). Silane Emergency Response, *Silane Safety Seminar - Safe Handling and Use of Silane Gas*, Taichung City, Taiwan, January 20, 2011.

14. Ngai, E. Y., Huang, K. P. P., Chen, J. R., Shen, C. C., Tsai, H. Y., Chen, S. K., Hu, S. C., Yeh, P. H., Liu, C. D., Chang, Y. Y., Peng, D. J., & Wu, H. C. (2007). Field tests of release, ignition and explosion from silane cylinder valves, *Process Safety Progress*, Vol. 26, No. 4, (December 2007), pp. 265-282, ISSN 1066-8527.

15. Rufin, D. (2011). Distribution Systems Best Practices, *Silane Safety Seminar - Safe Handling and Use of Silane Gas*, Taichung City, Taiwan, January 20, 2011.

16. Stockman, P. (2009). Going Green with On-Site Generated Fluorine: Sustainable Cleaning Agent for PECVD Processes, *EHS Workshop for Photovoltaic Manufacturers, Intersolar 2009* , Munich, Germany, May 28 2009.

17. Tsai, H. Y., Wang, S. W., Wu, S. Y., Chen, J. R., Ngai, E. Y. & Huang, K. P. P. (2010). Experimental Studies on the Ignition Behavior of Pure Silane Released into Air, *Journal of Loss Prevention in the Process Industries*, Vol. 23, No. 1 (January 2010), pp. 170-177, ISSN 0950-4230

CHAPTER 11

Transparent ZnO Electrode for Liquid Crystal Displays

Naoki Yamamoto, Hisao Makino and Tetsuya Yamamoto

Research Institute, Kochi University of Technology, Japan

1. INTRODUCTION

Recently, the scarcity and toxicity of indium, a major constituent element of ITO, has become a concern. Indium is a rare element that ranks 61st in abundance in the Earth's crust (Kempthorne & Myers 2007). In addition, the major amounts of indium consumed by the industries produceing the electronic devices such as liquid crystal displays (LCDs), touch- screens and solar cell systems are supplied by only a few countries. Furthermore, indium has also been suspected to induce lung disease, and particularly indium-related pulmonary fibrosis should be paid attention (Homma et al., 2005).

Transparent conductive oxides have become the focus of attention as a substitute material for ITO currently used for optically transparent electrodes in electronic devices. In particular, transparent conductive ZnO films are expected to be suitable materials to achieve such purposes because, in contrast with indium as a major constituent element of ITO, Zn is an element that the human body requires and is a component of some marketed beverages, in addition to having being used for years in cosmetics and as a vulcanization accelerator for rubber products such as tires. Furthermore, conductive and transparent ZnO films have low electrical resistance and high optical transmittance comparable with those of ITO films reported by some authors (Wakeham et al., 2009; Shin et al., 1999). We have developed the technology to form transparent conductive ZnO films with low resistance (2.4 □□□m for a 100 nm thick film (Yamada, et al., 2007)), optical transmittance exceeding 95% (film-only transmittance without that of the glass substrate) and high heat-resistance (thermally stable until 300-450 °C (Yamamoto, N. et al., 2010)). The technology of transparent conductive ZnO films applied as alternatives to ITO electrodes for LCD panels is described in this chapter.

2. PREPARATION OF TRANSPARENT AND CONDUCTIVE ZNO FILM

Ga-doped ZnO (GZO) and Al-doped (AZO) films have been widely studied as the most promising transparent conductive films as alternatives to ITO films used in electronics devices such as LCDs, LEDs and solar cells.

2.1 Magnetron sputtering system

Conventional magnetron sputtering systems, planar- and cylindrical-types (Carousel-type), were used to form transparent ZnO thin films. A schematic diagram of the cylindrical-type magnetron sputtering system is shown in Fig. 1 (a). In the cylindrical-type magnetron sputtering system, a drum with samples set on its surface is rotated concentrically in the chamber with the sputtering target set on the inside wall. A film can be formed by sputtering with dc power (noted as dc MS) and radio frequency power combined with dc power (noted as rf+dc MS) applied to the sputtering target.

2.2 Reactive plasma deposition system

Figure 1 (b) shows a schematic diagram of the reactive plasma deposition system (RPD) (Yamamoto, T. et al., 2008)., which is a type of ion-plating method. An Ar plasma stream is generated by a pressure gradient arc plasma source (Uramoto gun) at the cathode is introduced by control of the electric and the magnetic field to the evaporation source tablet inset in the hearth at the anode. The particles evaporated from the source are deposited onto the substrate set on the tray traveling in front of the heater.

The specifications for the formation of conductive transparent ZnO film using the magnetron sputtering systems and the RPD system are summarized in Table 1

An Ar plasma stream is generated by a pressure gradient arc plasma source (Uramoto gun) at the cathode is introduced by control of the electric and the magnetic field to the evaporation source tablet inset in the hearth at the anode. The particles evaporated from the source are deposited onto the substrate set on the tray traveling in front of the heater.

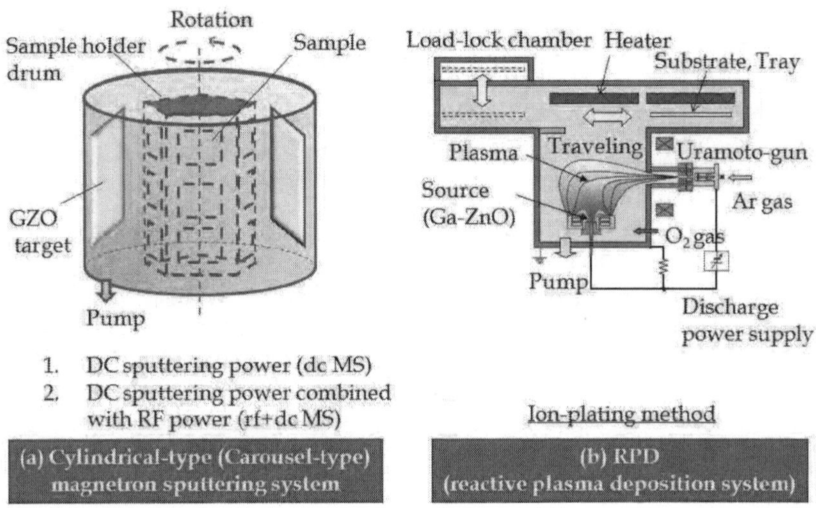

Figure 1. Schematic diagrams of the deposition systems for the transparent conductive ZnO films.

Table 1. Specifications for the formation of GZO or AZO films (Yamamoto. N. et al., 2011a & 2011c).

	Magnetron Sputtering (MS)		RPD
	dc MS	rf+dc MS	
Ga_2O_3/Al_2O_3 content in ZnO source (wt%)	Ga_2O_3: 3.0 - 6.0 Al_2O_3: 2.0 - 5.0		Ga_2O_3: 3.0 - 5.0
Power (kW)	0.1 - 2.0	rf: 0.1 - 1.5, dc: 0.1 - 1.5 rf/dc = 0.5 - 2.0	discharge current: 140 - 150 (A)
Operation pressure (Pa)	0.1 - 0.8	0.1 - 0.8	0.4 - 0.6
Operation temperature (°C)	25 - 350	25 - 350	25 - 250

The specifications for the formation of conductive transparent ZnO film using the magnetron sputtering systems and the RPD system are summarized in Table 1

3. BASIC CHARACTERISTICS OF TRANSPARENT CONDUCTIVE ZNO FILM

The fundamental characteristics of transparent conductive ZnO films for application to LCD panels are discussed in this section.

3.1 Crystalline structure of transparent conductive ZnO film

X-ray diffraction (XRD; ATX-G, Rigaku) and transmission electron microscopy (TEM; H- 9000UHR; Hitachi High-technologies Co.) were applied for analysis of the crystalline structures of transparent conductive ZnO films.

The crystalline structures and orientations of the GZO films were analyzed using both out- of-plane XRD (widely used X-ray diffraction analysis) and in-plane XRD (grazing-incidence wave-dispersive X-ray analysis (Ofuji et al., 2002)). For measurement using the in-plane XRD technique, a Cu Kα X-ray beam with a wavelength of 0.154184 nm was irradiated at a low angle of incidence to the surface of the sample (0.35°). The incident angle is close to the total reflection angle of X-ray for ZnO.

XRD patterns obtained from the GZO films deposited by dc MS, rf+dc MS or RPD were almost identical and had the wurtzite crystalline structure, as with the ZnO films. A typical XRD pattern obtained from a GZO film is shown in Fig. 2.

The in-plane XRD diffraction pattern shows that no diffraction peaks from the ZnO(00x) crystal planes were evident (Fig. 2(a)). In contrast, the out-of plane XRD pattern shows only the (002) and (004) diffraction peaks of the GZO film (Fig. 2(b)).

The appearance of these diffraction peaks clarified that (1) the GZO polycrystalline film consists of the wurtzite structure. (2) the *c*-axes of the wurtzite structure coincides with the direction normal to the GZO film surface, and (3) the *a*-axes of the wurtzite cell structure coincides with the direction in the plane of the film. The TEM image in Fig. 3(a) and the cell structure shown in Fig. 3(b) explains the structure. Columnar grains comprise the interior of the polycrystalline GZO films (Yamamoto. N. et al., 2008). Such crystalline structures also appeared in films formed in the temperature range of 150-250 °C using dc MS, rf+dc MS and RPD.

The lattice constants for the *c*- and *a*-axes, and the volume of the wurtzite crystalline unit cell in 100 nm thick GZO films prepared at 180 °C by dc MS, rf+dc MS and RPD were derived using the XRD peaks diffracted from the (00x) and (x00) crystalline planes and are compared in Fig. 4 (Yamamoto, N. et al., 2010). The lattice constants of the GZO films prepared by RPD were shorter than those of the films formed by magnetron sputtering (Fig. 4(a)). The *c*-axis of the rf+dc MS film was especially expanded toward the direction normal to the surface of the substrate compared with the other films. The *a*-axis was also expanded toward in the direction of the plane of the film. As a result, the cell volume of the wurtzite structure in the films prepared by rf+dc MS were larger than those formed by dc MS and RPD, as shown in Fig. 4 (b).

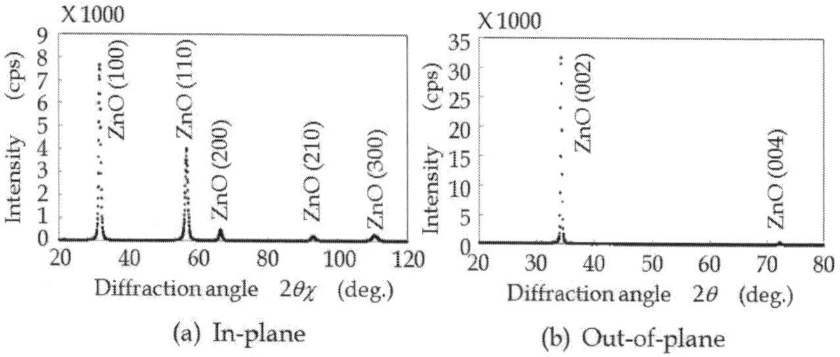

Figure 2. Typical XRD profile of a 150 nm thick GZO film prepared at 180 °C using dc MS.

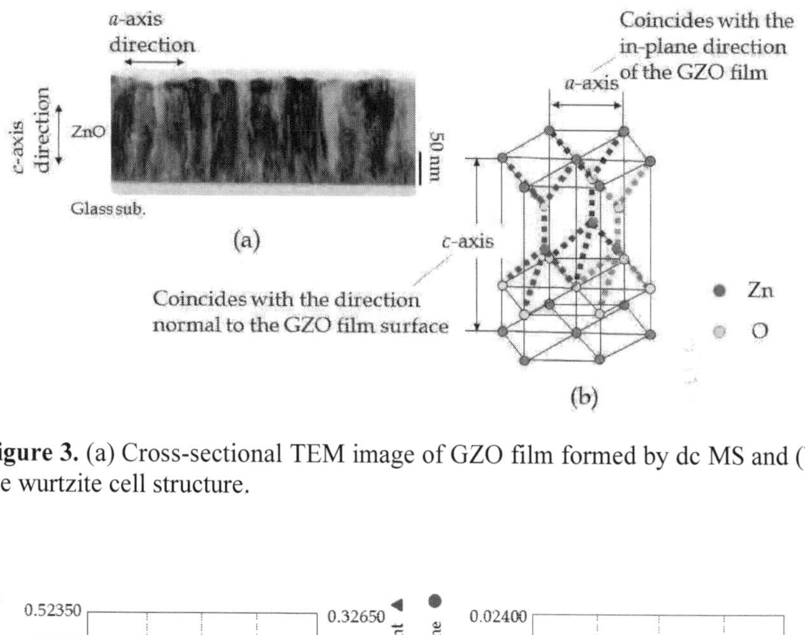

Figure 3. (a) Cross-sectional TEM image of GZO film formed by dc MS and (b) the wurtzite cell structure.

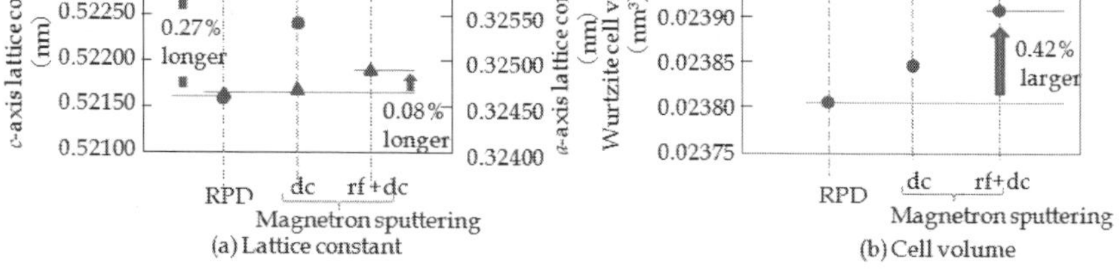

Figure 4. Comparison of the *a*-axis, *c*-axis lattice constants and the unit cell volumes in crystalline ZnO-based wurtzite structures of films prepared using dc MS, rf+dc MS and RPD (Yamamoto,N. et al 2010).

3.2 Resistance of transparent conductive ZnO film

The resistivity of the film is one of the most important characteristics required for application of alternative transparent electrodes to ITO used in LCDs. The resistivity of ITO transparent films is in the range of 1.3–3.8 □□m (Wakeham, et al.. 2009; Shin et al., 1999). Therefore, transparent conductive ZnO films as an alternative must have a resistivity of less than ca. 3.8 □□m.

In the case of thin films, the resistivity is generally derived from the carrier-flow in the plane of the films. The resistivities of even metal films increase with the decrease in film thickness, especially less than ca. 100 nm. Such a phenomenon is caused by the increase in the frequency of collisions or scattering with the carrier–flow and the film surface, the interface with the substrate and irregular crystalline structures in the region of the substrate. The resistivity of the GZO film also shows a similar dependency on the film thickness, as shown in Fig. 5(a).

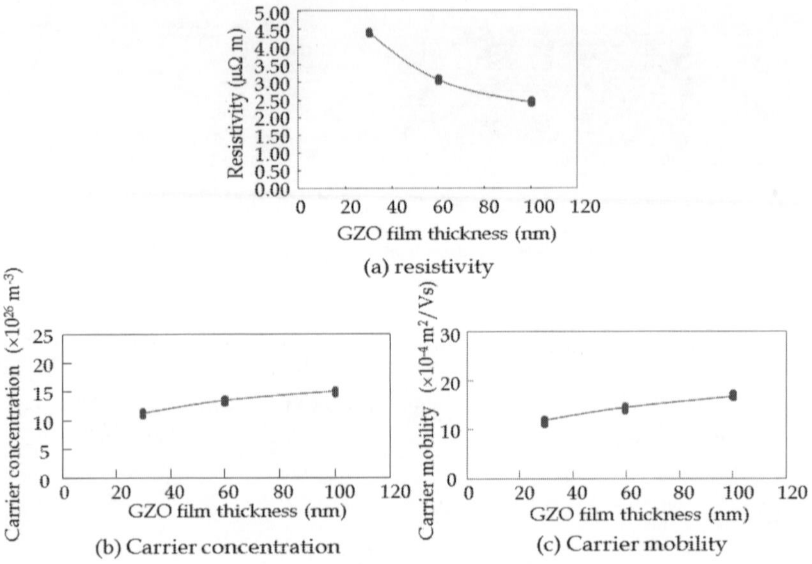

Figure 5. Electrical characteristics of GZO film as a function of the film thickness.

As the film thickness decreases, the carrier mobility and concentration in the film decreases, as shown in Figs. 5(b) and (c). The data in Fig. 5 was obtained from GZO films prepared using RPD at 180 °C. The electrical characteristics of transparent conductive ZnO films formed by the magnetron sputtering showed similar dependencies on the film thickness, although the values were significantly affected by the formation conditions, i.e., type of dopant and its concentration, the deposition equipment, temperature, pressure and the electrical power supplied to the source during deposition.

Figure 6 shows a comparison of the resistivities of GZO and AZO films formed with an Ar sputtering pressure of 0.66 Pa at room temperature or at 180 °C using the conventional planar magnetron sputtering system. The ZnO sputtering target materials for the deposition of GZO or AZO contained 4 wt% Ga2O3 or 2 wt% Al2O3, respectively. The total sputtering power used in both cases of dc MS and rf+dc MS was set to 200 W.

The resistivities of the AZO films were 1.5 times or more higher than those of the GZO films. The resistivities of the films formed by rf+dc MS were lower than those of films formed by dc MS (Yamamoto, N. et al., 2011a & 2011c). However, this relationship does not correspond to the lattice constants and cell volumes of the wurtzite structures of the GZO films formed by rf+dc MS, which were larger than those of the films formed by dc MS, as shown in Figs. 4(a) and (b). It is generally considered that the probability of carrier hopping among atoms decreases with distance from each atomic site, and as a result, the resistivity of the film decreases with the length of the lattice constant. The reason for the contradiction between the resistivity and the lattice constant (length) is yet to be clarified.

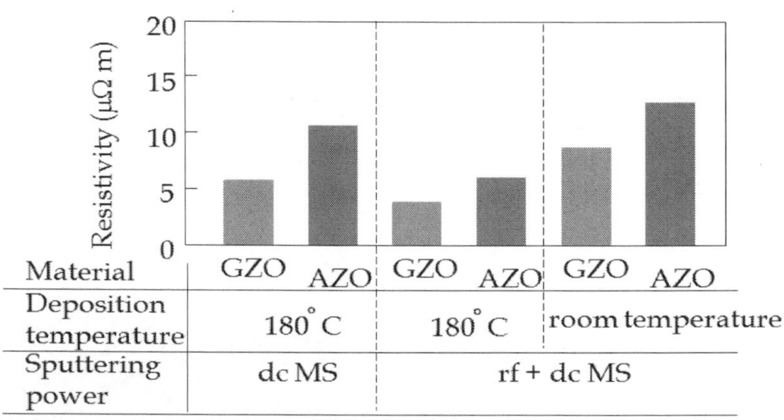

Figure. 6. Comparison of resistivity for ca. 150 nm thick transparent ZnO films formed by dc MS and rf+dc MS

3.3 Visible light transmittance of transparent conductive ZnO film

Alternative materials to the ITO electrodes used for LCDs are required to have transmittance for visible light comparable with that of ITO films. The optical transmittance of the films was analyzed using an optical spectrophotometer (U-

4100 UV-Visible-NIR spectrophotometer, Hitachi High-tech. Co. Ltd.). The transmittance of the GZO films in the wavelength range of ultraviolet (UV) to near-infrared (NIR) is shown in Fig. 7(a) The transmittance values presented in this section include the transmittance of the glass substrate. The absorption edges in the spectra are shifted slightly to the shorter side from the wavelength (ca. 370 nm) corresponding to the bandgap of undoped-ZnO according to the Burstein–Moss effect (Moss 1980).

On the other hand, the transmittance in the NIR region was significantly reduced from an increase of reflectance due to the plasma resonance of electron gas in the conduction band in the films with highly density carriers (electrons) of $8\square10^{26}$ to $1.5\square10^{27}$ m^{-3} (Jin et al., 1988; Dong & Fang 2007). The transmittance spectra in the wavelength region of visible light are scaled up in Fig. 7(b). Undulations in the transmittance spectra as a function of wavelength appeared due to optical interference phenomena caused at the surface and at the interface with the glass substrate because the dependency of the 110 nm thick sample on the wavelength differed from those of the ca. 150 nm thick samples. The maximum, minimum and average transmittance of each sample in the range of 400–800 nm were 90.5 – 91.4, 73.1 – 86.4, 87.2 – 87.8, respectively.

The transmittance of a 150 nm thick GZO film formed at 200 °C by conventional dc MS on a 1.1 mm thick Corning #1737 glass substrate was compared with polycrystalline and amorphous ITO films with ca. 150nm thickness formed using a similar sputtering system, and the results are shown in Fig. 8 (Yamamoto, N. et al., 2011a & 2011c). The exact deposition conditions is proprietary and therefore limited to announcement by Geomatec Co., Ltd. Japan. The amorphous and polycrystalline ITO films were formed at approximately 50-70 and 250-300 °C, respectively. The optical transmittance of the GZO film surpassed those of the ITO films in the entire wavelength range of visible light. Visible light in the range of 400-550 nm was transmitted through the GZO film with 2-7% higher transmittance than those of the ITO films.

3.4 Residual stress in transparent conductive ZnO film

The development of transparent conductive films that are resistant against external forces and forces induced by thermal processes (maximum temperature: ca. 250 °C) during LCD fabrication requires investigation of the fundamental mechanical characteristics of the films, such as residual stress, thermal stress, strain, Young's modulus, coefficient of thermal expansion, adhesive force and creep. In this section, the residual stress of GZO films is discussed. The conventional optical lever method was applied for residual stress analysis using a HeNe-laser beam at 633 nm (F2300, Flexus Co.). The Si wafer substrate was used as a reflector for the laser beam. The substrate with film was placed on the three projections attached with an equilateral-triangular layout on the surface of the heater plate. The radius of curvature of the Si wafer was measured by detecting the reflected beam. The stress in the film was then derived using the curvature radius, the linear coefficient of thermal expansion (3.34×10^{-6} K^{-1}) and the Young's modulus (160 GPa) of the Si substrate (Wortman & Evans 1965).

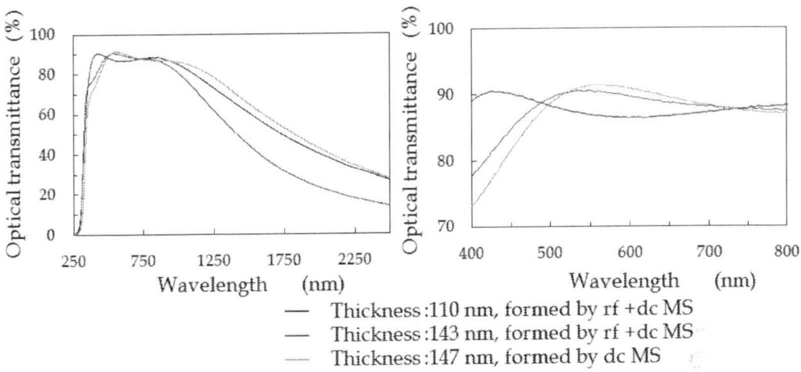

(a) Spectra in the range from UV- to near infrared-wavelength

(b) spectra in the 400 nm -800 nm range of the visible light.

Figure. 7. Comparison of optical transmittance of films formed at 180 °C using rf+dc MS and dc MS on 0.7 mm thick Corning #1737 glass substrates.

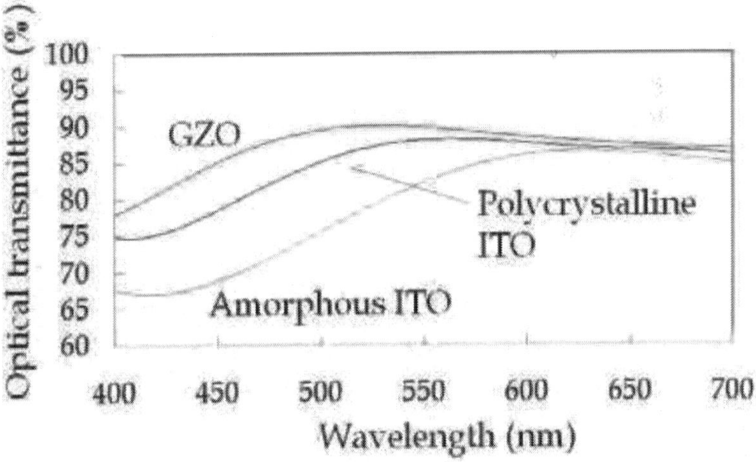

Figure 8. Comparison of the optical transmittance of transparent GZO and ITO films.

The wafer was heated from 25 to 500 °C and subsequently cooled to 25 °C for heat-cycle testing of the film, and the curvature radius of the Si wafer was measured *in situ* during testing. The residual stresses of all the as-deposited films were artificially set to compressive, according to the controlled deposition processing factors. The residual stress in a film is dependent on the degree of energetic particle bombardment, that is, energy striking the condensing film during deposition by magnetron sputtering or RPD with plasma discharge discharge (Thornton, & Hoffman, 1977; Yamamoto, N. et al., 1986).

The residual stress properties of the films were clarified by applying heat-cycle testing. Heat-cycle testings was applied twice to each sample. The samples were heated at a rate of 2.8 °C/min until thermal equilibrium (thermally quasi-static condition) and *in situ* measurements of residual stresses in the films were conducted during heat-cycle testing. The residual stresses of the films formed at 180 °C using rf+dc MS, dc MS and RPD had following similar behaviors, as shown in Fig. 9 (Yamamoto, N. et al., 2008 & 2010). (I) the residual stress in the films was significantly reduced in the range from 200 to 400 °C during the first heating-up process (step 1); there was an increasing tendency toward strongly compressive residual stress, which changed abruptly to the tensile direction upon heating. This critical temperature of change is dependent on the deposition method of the films. (II) the stress decreased monotonically with cooling from 500 °C (step 2), and (III) the dependence of the stress in each film on the temperature during the second heat-cycle (steps (3) and (4)) almost coincided with that in step (2). These phenomena present evidence that the extrinsic stress components in each film are removed by annealing during the first heating step to 500 °C, and the resulting stress is composed of only thermal stress, i.e., intrinsic stress. The thermal stress is caused by the difference in the linear coefficient of thermal expansion between the GZO film and the Si substrate.

On comparatively thicker GZO films (ca. 500 nm in Fig. 9) formed by dc MS and RPD, the temperature dependencies in step (1) were closer to those in the second cycle testing. The main component of internal stress in such thick films was thermal stress (intrinsic stress), even before heat-cycle testing of the as-deposited film. The GZO films approached the ideal crystalline structure as the thickness increased. The overall temperature dependency change of the residual stresses according with the film thickness corresponds approximately with the increase in crystalline irregularity with the distance from the interface to the substrate. On the other hand, the GZO film thicker than 500 nm formed by rf+dc MS did not provide the same stress behavior in step (1) in the GZO closed to those in steps (2) to (4). High strain or irregular crystalline structures were present until a far distance from the substrate surface of the film.

4. FABRICATION OF LCD PANELS WITH TRANSPARENT ZNO ELECTRODES ON RGB COLOR FILTERS

4.1 Intrinsic weakness of transparent conductive ZnO film in the LCD manufacturing environment

The fundamental properties of GZO films as an alternative to ITO transparent electrodes have been clarified with experimental results. However, the properties required for application in the production of the LCDs have not yet been discussed thoroughly. ZnO or materials based on ZnO have significant impediments to LCD applications. Such materials have amphoteric properties, so that their films have the extremely low tolerability to the various process conditions and environments used for manufacturing and operating LCD panels, i.e., acidic, basic and highly humid environments. Furthermore, Zn atoms are easily volatilized from materials by heating processes at relatively low temperature such as 300- 400 °C. Transparent electrode films are required to endure manufacturing processes at ca. 250 °C for the fabrication of LCDs. A diagrammatic cross-sectional illustration of an LCD is shown in Fig. 10.

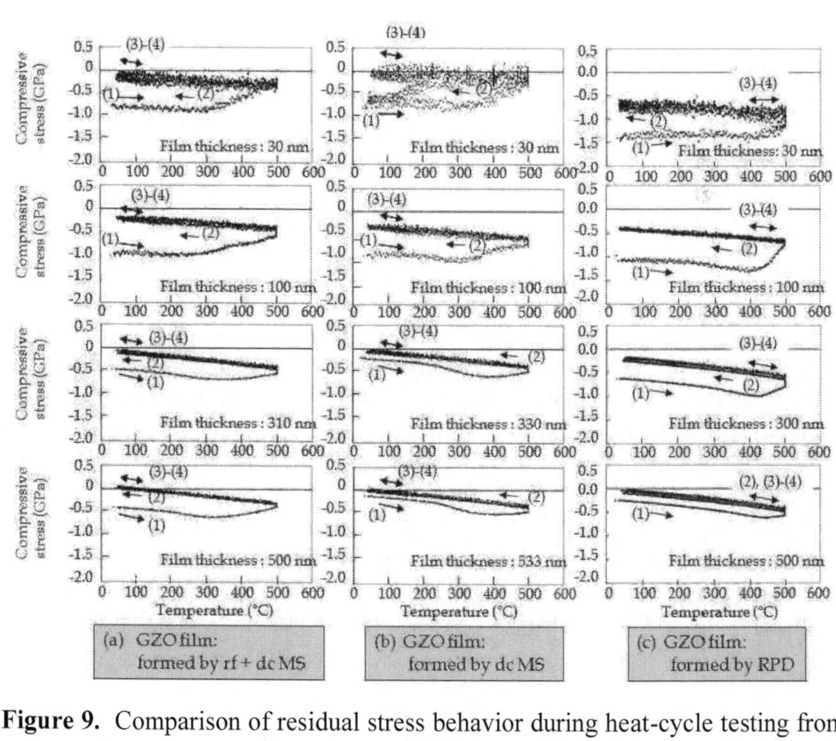

Figure 9. Comparison of residual stress behavior during heat-cycle testing from room temperature to 500°C in an Ar atmosphere.

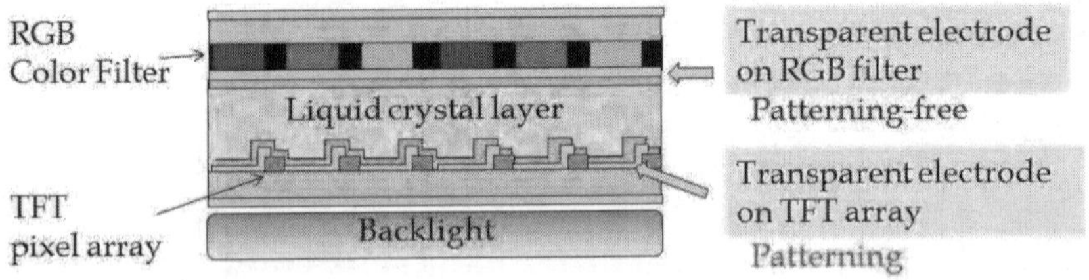

Figure 10. Schematic cross-sectional view of an LCD.

The transparent conductive films on the RGB color filter is generally used without patterning. On the other hand, the transparent electrode on the TFT pixel array is formed by patterning the film using wet-chemical etching techniques. Therefore, as the first stage of transparent ZnO electrode development, the GZO films were applied as only the transparent electrode on the RGB color filter in an LCD. In this case, the conventional ITO electrode was used on the TFT pixel array. Concurrently, a technique to form the fine-line patterns of the transparent conductive ZnO film was developed using wet-chemical etching.

4.2 Heat resistance against thermal processing during LCD fabrication

Figure 9 should be referred to again for discussion regarding the heat resistance of GZO films, especially with regards to the behavior of 100 nm thick GZO films because this thickness is close to that used for transparent electrodes in LCDs. During the first heating (step (1)) in the heating cycle test, the compressive residual stress started to reduce suddenly at a specific temperature. The temperature range in which a detectable reduction in the compressive stress occurred in each type of film formed by rf+dc MS, dc MS and RPD was 200-250, 250-300 and 350-400 °C, respectively.

The cause of the decrease in the residual stress with heating to higher than the critical temperature was analyzed using thermal desorption spectrometry (TDS; EMD- WA1000S/W, ESCO Co., Ltd.). The dependence of the amount of Zn mass fragment in the TDS mass spectra on the temperature had the highest correlation with stress for various materials subliming from each GZO film, as shown in Fig. 11(a) (Yamamoto. N. et al., 2008). The temperature at which an observable change in the amount of Zn in the three film types due to volatilization was at 200-250, 250-300 and 300-400 °C, respectively (indicated with arrows). Therefore, it was concluded that the volatilization of Zn from the films at these temperatures during step 1 in the heat-cycle test caused the decrease in the compressive residual stress. The correlation indicates that shrinkage of the film volume accompanied the volatilization of Zn, which resulted in a reverse of the temperature-dependence of the residual stress from increasing compressive stress toward tensile stress.

The order among the critical temperatures at which the change of residual stress and the amount of volatilized Zn began to become significant for the three film types has a good correlation with the order among the lengths of the lattice constant or the wurtzite cell volume as shown in Fig. 4. Longer distances between the atoms in the wurtzite structure is likely to cause weakening of the atomic binding forces, which results in lowering of the heat-resistance of the material.

The heat-resistant properties of the three film types derived from these experimental results were in following order: rf+dc MS < dc MS < RPD. The order of the heat-resistance with respect to the electrical characteristics was exactly the same as that with respect to Zn volatilization and the residual stress. The change in the resistivity, carrier concentration and carrier mobility caused by annealing (heat cycle testing) are compared for films formed by rf+dc MS and RPD in Fig. 11(b). The critical temperatures estimated from the electrical characteristics for the three film types were 200-250 °C (rf+dc MS) < 250-300 °C (dc MS) < 350- 400 °C (RPD). This order corresponded exactly with the release of compressive residual stress and Zn volatilization in step (1) during the heat-cycle testing (Yamamoto, N. te al 2010).

Therefore, the films formed by dc MS and RPD have heat resistance properties that are suitable for use as an ITO substitute material for optically transparent electrodes in LCDs. In the case of the film formed by rf+dc MS, the heat resistance characteristics were on the borderline for LCD application. It would be necessary to improve the properties of the rf+dc MS film by some process.

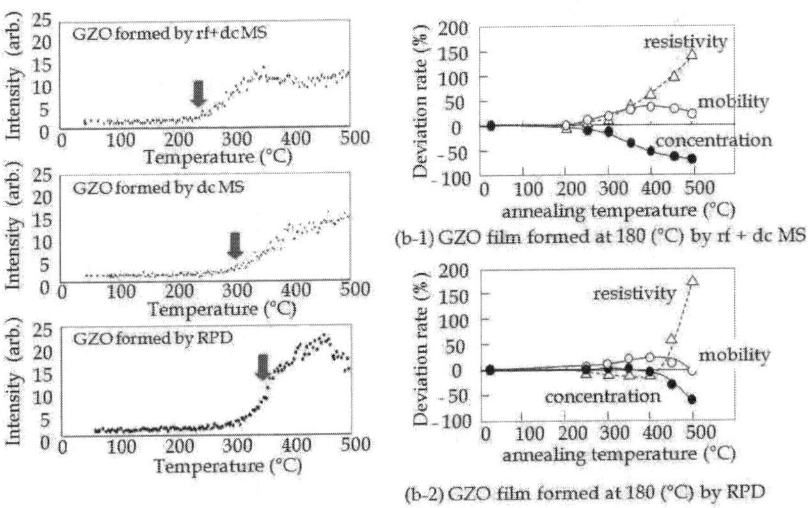

(a) TDS spectra (b) Electrical characteristics

Figure 11. (a) Comparison of TDS mass spectra for the volatilization of Zn plotted as a function of temperature for 100 nm thick GZO films formed by rf+dc MS, dc MS and RPD at 180 °C.

(b) Degradation rates of resistivities, carrier concentrations and carrier mobilities in the films formed by rf+dc MS and RPD at 180 °C as a function of temperature when annealed in flowing Ar (Yamamoto, N. et al., 2010).

5. FABRICATION OF LCDS WITH TRANSPARENT ZNO ELECTRODE

As the first stage of transparent ZnO electrode development, the GZO electrodes were only applied on the RGB color filter for the manufacture of the LCDs. Conventional ITO films were used as the electrode on the TFT pixel arrays in the LCDs. Among the three film types, the films formed by RPD have the best suited characteristics, such as low electrical resistance and high heat-resistance. However, the current RPD system is not in the usable field for the large size motherglass (1500×1800 mm^2, 2160×2400 mm^2 and 2850×3050 mm^2) used in the production lines of large size LCD TVs, such 6G (generation), 8G and 10G. Therefore, the conventional dc MS technique, which is used for the formation of ITO transparent electrodes used in commercially available LCDs, was used for the fabrication of LCDs with the GZO electrodes on RGB color filters.

150 nm thick transparent GZO films were formed on the RGB color filters at 150 °C using dc MS. The LCDs were fabricated without change to the process flow, except for the GZO deposition. The fabrication process flow is generally kept an industrial secret. Therefore, the order of the most basic process steps for manufacturing the LCDs are shown as a reference in Fig. 12 (Yamamoto, N. et al., 2010).

The fabrication of the RGB color filter side module and the TFT pixel array side module were carried out separately from each other. After the fabrication reached the final step in each side-module process flow, the two side modules are combined with each other. Consequently, the liquid crystal material is poured into the gap between both substrate sides. LCD fabrication is accomplished through hinging the bezel and attaching various controlling circuits. Finally the completed LCDs are tested for long term at 50-70 °C in a high humidity (90-95%) environment.

Figure 13 shows the typical display of 3 inch size LCDs and the motherglass before cutting each 3 inch panel part (Yamamoto, N. et al., 2010). The production procedure and process conditions were the same as those for commercially available TFT-LCDs with conventional ITO transparent electrodes, except for the processing step for the formation of transparent conductive GZO films.

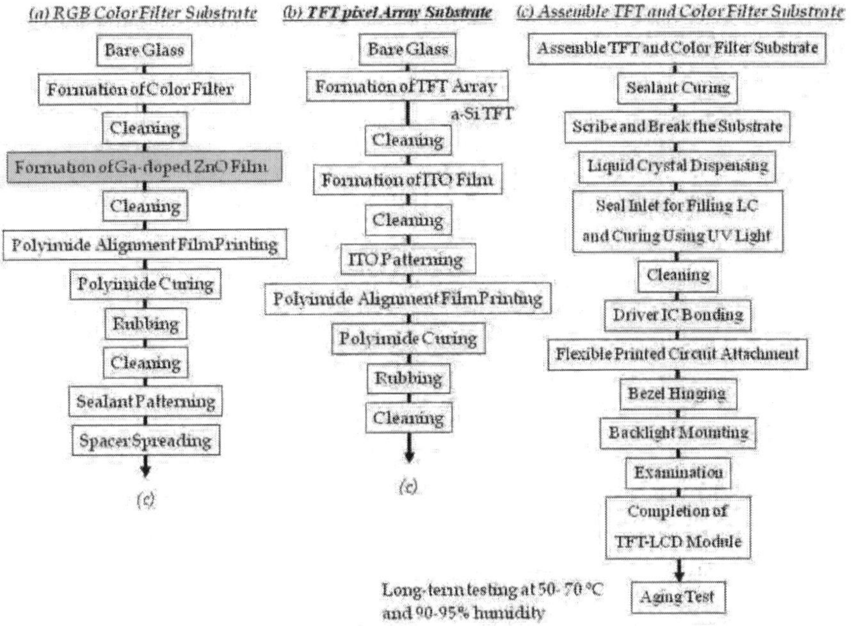

Figure 12. Basic process flow for manufacture of an LCD with GZO on the RGB color filters.

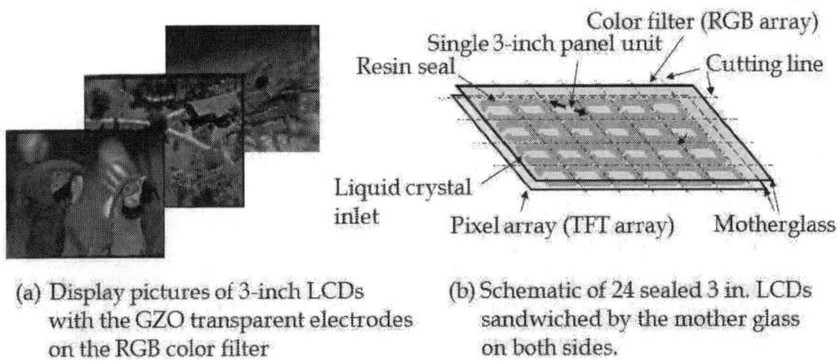

Figure 13. (a) Display pictures of 3 inch LCDs with transparent GZO electrodes on the RGB color filters. (b) Schematic of 24 sealed 3 inch LCD sandwiched by the motherglass on both sides.

The displays of 20 inch size LCD TVs with GZO transparent electrodes on the RGB color filters are compared with those of commercially available LCD TVs with ITO electrodes in Fig. 14 (Yamamoto, N. et al., 2011a & 2011c).

The various display properties, such as the contrast ratio of the module, and the chromaticity diagram characteristics, of the LCD TVs with the GZO electrodes were equivalent to those of the LCD TVs with conventional ITO electrodes. The average visible light transmittance of the TV module with the GZO transparent electrode exceeded that of the conventional TV modules with ITO electrodes by 4–5%. These superior characteristics of the GZO module were confirmed by the experimentally measured transmittance of the GZO film in the short wavelength (blue) region of visible light, which exceeded that of polycrystalline and amorphous ITO films (Fig. 8).

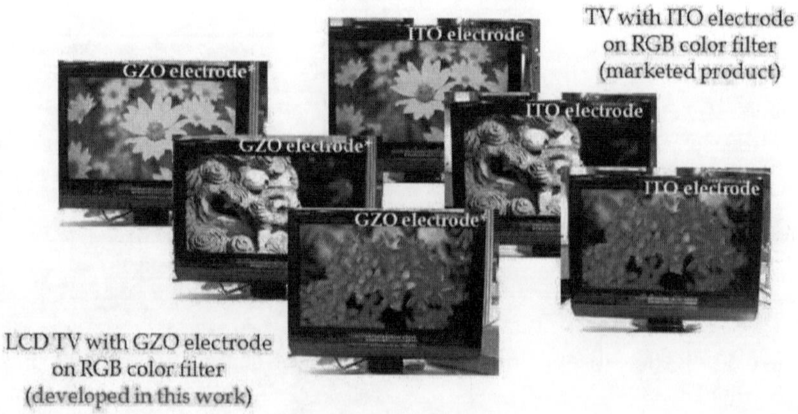

Figure 14. Comparison of the display pictures between 20 inch LCD TVs with GZO and ITO electrodes on the RGB color filter.

The major difference between the 3 inch size LCDs and the 20 inch LCD TVs are the spacers set at the inter-gaps between both side modules. Bead type spacers are distributed between the modules to form a space to inject the liquid crystal, as shown in Fig. 13(b). In the case of the 20 inch LCD TVs, column (rib) type spacers were formed on the plates of the RGB color filters using a conventional photolithography patterning technique. The process steps and the detailed flow for the manufacture of LCD TVs does deviate from that presented in Fig. 12; however, this is proprietary knowledge and the intellectual property of each company.

The display performance of the completed 3 inch LCDs and the 20 inch LCD TVs did not degrade even after long-term operating tests for over 1000 h (on gong) at 50-65 °C in a high humidity (90-95%) environment. Prior to the long-term operation testing, there was some concern regarding degradation of the display performance due to the high humidity environment because the resistance and transmittance of the transparent ZnO film deteriorates under high humidity (Nakagawa, et al., 2009). However, the long-term operation testing results for the LCDs confirmed that transparent ZnO electrodes sealed in the modules of LCDs were not affected by the external environment.

6. FORMATION OF TRANSPARENT ZNO ELECTRODES WITH FINE-LINE PATTERNS ON TFT PIXEL ARRAYS

Transparent films formed on the TFT pixel array must be finely patterned using wet- chemical etching techniques. GZO is amphoteric and has low resistance to such agents; therefore, there are no reports on the formation of transparent ZnO patterns with line- widths finer than approximately 10 □m. Thus, the fine-patterning of GZO thin films by wet- chemical etching is a significant developmental challenge.

Fine-patterns of transparent conductive ZnO films were formed by the process flow shown in Fig. 15 (Yamamoto, N. et al., 2011a; 2011b & 2011c). At first, a positive-type novolac- photoresist layer was formed on a transparent conductive ZnO film using a spin-coating applicator for the formation of fine-patterned films. A contact aligner (UV-light exposure system) was used to print the designed mask patterns onto the photoresist layer. The UV- exposed parts in the photoresist layer were removed with a photoresist-developer containing tetramethyl ammonium hydroxide (TMAH, (CH3)4NOH). The unexposed parts remained as photoresist patterns. Patterning of the GZO films was conducted using a wet- chemical etching technique with organic-acidic etchants; carboxylic acid agents with the photoresist-patterns as the etching mask. The patterning-mask photoresist remaining on the ZnO material patterns was subsequently removed using a photoresist-stripper containing an amine (ELM-R10-F22, produced by Mitsubishi Gas Chemical (MGC) Co., Inc.). The key to realize fine-patterns with widths of a few micrometers is fulfilled by the following three requirements. (1) Development or choice of a suitable developer for the photoresist, i.e., alkaline solution. An aqueous solution prepared by the addition of TMAH to deionized pure water was used as the developer. (2) Development or choice of an acidic solution (etchant) for wet-etching (patterning) of the transparent conductive ZnO films. Solutions based on organic acids and inorganic acid with different pH values were prepared in this work. (3) Optimization of the photolithography and etching processes for GZO films.

Fine-patterns with line and space widths of a few micrometers were successfully formed using this technology, as shown in the optical micrographs in Fig. 16 (Yamamoto, N. et al., 2011a; 2011b & 2011c).

Micrographs (a) and (b) show typical meander patterns with line and space widths of 5 □m and comb-like line and space widths of 3 □m, respectively.

Figure 16(c) and (d) shows dense 2 μm line and space patterns of 120 nm- and 150nm- thick GZO film formed successfully using the developed wet-chemical etching technique.

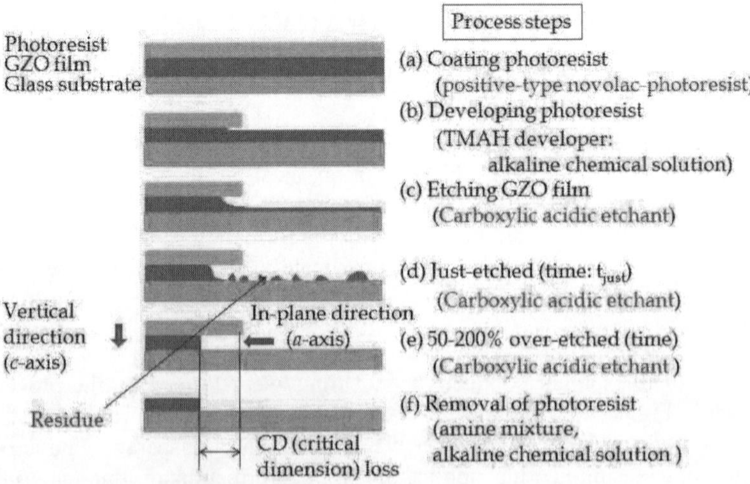

Figure 15. Process flow for patterning GZO films using wet-chemical etching techniques

The technology for fabricating fine-patterns of transparent conductive ZnO films developed in this work is summarized as follows. The key factors are (1) the development or selection of appropriate chemical agents, and (2) the determination of an appropriate pH range for each agent (Yamamoto, N. et al., 2011a; 2011b & 2011c).

1. Alkaline chemicals for photolithography process
   a. Developer: TMAH aqueous solution pH: 12.0-13.0
   b. Stripper: Amine chemical solution pH: 11.0-12.0 (ELM-R10-F22, MGC Inc.)
2. Acidic agent for wet-chemical etching (patterning) pH: 5.5-6.8
   a. Organic acidic etchants (carboxylic acid series) prepared by MGC Inc. (Chemical composition: nondisclosur. Not for sale at the present stage)
   b. Inorgabic acidic etchants prepared by Naoki Yamamoto
3. Optimization and control of conditions related to lithography process (UV light exposure, baking of photoresist) and etching process (temperature of etchant, rinsing and drying after etching)

It should be noted here that the patterns shown in Fig. 16 had smooth edges and no residues were observed on the glass substrates after etching of the GZO films. In addition, the 2 □m width line patterns of the GZO film are comparable to the narrowest ITO transparent electrodes formed by wet-etching techniques using leading-edge proximity exposure systems available in commercial LCD production lines.

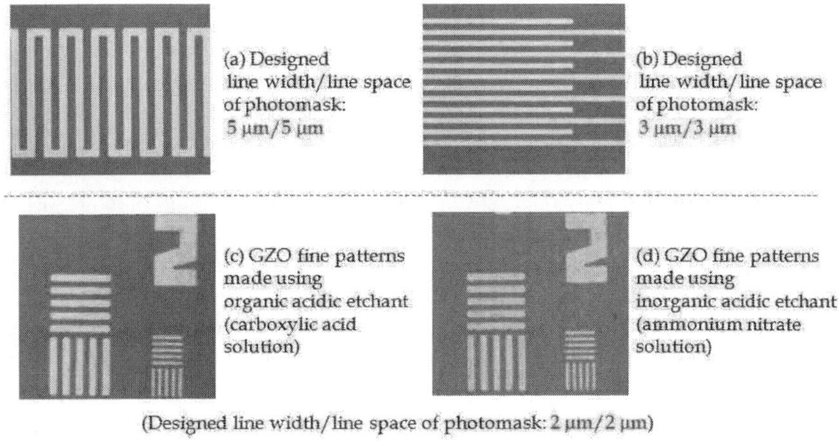

Figure 16. Typical GZO patterns formed using a weakly acidic etchant (pH 5.5 – 6.8).

7. SUMMARY

GZO and AZO films are considered to be the most suitable conductive ZnO films for the current purpose of LCD fabrication. Particular focus was made on the GZO films, because the resistivities of GZO films are lower than those of AZO films.

GZO films have the wurtzite structure as with ZnO films. The resistivity of the GZO films reached approximately 2.4 □□m, which is in the range of 1.7–3.8 □□m reported for ITO films formed at temperatures lower than 250 °C.

Conductive ZnO films are amphoteric and are therefore susceptible to acidic and alkaline environments. Therefore, the most important issue for the development of transparent GZO electrodes for LCDs is the development of wet-chemical etching techniques to form fine- patterns with widths of a few micrometers.

The fabrication of LCDs includes process steps that require heating up at 200-250 °C. The thermal resistance of GZO films was investigated from the residual stresses behavior of the films and the volatilization of Zn in the films during heat-cycle testing between room temperature and 500 °C. During the first heating step, the critical temperature where significant Zn volatilization from the GZO film begins coincided with the temperature where significant changes begin to appear in the residual stress and electrical characteristics, that is, resistivity, carrier mobility and carrier concentration. The relationship of these changes with the volatilization of Zn caused by heating the films was clarified. The temperature at which significant change occurs is the critical temperature of the GZO film. The critical temperature was dependent on the method of film formation as follows. The critical temperatures of the films were in the order of 250 °C (rf+dc MS film) < 300 °C (dc MS film) < 400 °C (RPD film). Films formed using dc MS and RPD would therefore be resistant to the thermal process steps conducted at a maximum temperature of 250 °C for the manufacture of LCD panels (TV sets). The transmittance of the GZO film exceeded that of the conventional ITO electrode by ca. 2-7% in the short wavelength region of visible light. As the first development stage of the transparent ZnO electrode, ca. 150 nm thick GZO films were used as only the transparent electrodes on the RGB color filters in 3 inch LCDs and 20 inch LCD TVs. Conventional ITO electrodes were used as the transparent electrodes on the TFT pixel arrays. The initial display performance of both the 3 inch LCDs and the 20 inch LCD TVs was maintained even after long-term operation testing (>2000 h) at 50-65 °C in a high humidity (90-95%) environment. A wet-chemical etching technique was developed concurrently for the formation of fine-line dense patterns with widths of a few micrometer using the amphoteric GZO films as the 2nd stage of transparent ZnO electrode development. This involved (1) development or selection of appropriate alkaline chemicals for photolithographic processing, (2) development of acidic chemicals for the etching of transparent ZnO films, and (3) optimization of the photolithography and etching processes. The suitable pH range of each chemical agent was determined experimentally, and included the following: (1) TMAH photoresist-developer: pH 12.0–13.2, (2) an etchant for patterning the transparent ZnO film: pH 5.1-6.8, and (3) photoresist remover (stripper): pH 11.0–12.0. Fine line dense patterns with 2 □m line/2 □m space widths were successfully fabricated from 50–150 nm GZO films using these selected reagents and the developed technique. AZO fine-line patterns with the same widths could be also formed using the pattering techniques, because the CD loss (the critical dimension loss is defined as the distance encroached in the ZnO conductive films by acidic etchant from the edge of the patterns under the photoresist layer. reffer Fig.15) dependency rates of AZO films on the over-etched times processed using the acidic etchant were the same as those of the GZO films.

8. FUTURE PERSPECTIVE OF TRANSPARENT ZNO ELECTRODE TECHNOLOGY

It is expected that LCD TVs manufactured in near future will have the transparent ITO electrodes replaced with transparent ZnO electrodes. Furthermore, the successful development of fine-patterning technology for transparent ZnO films should realize the application of transparent ZnO electrodes to not only LCD TVs that are operated by the VA (vertical alignment) method, but also LCD TVs driven by the IPS (in-plane switching) method. In addition, it is expected that this fine-patterning technology will enable the application of transparent ZnO electrodes to not only LCDs, but also organic LEDs (OLEDs), touch panels LEDs and solar batteries. This technology may also contribute to the development of discrete electronic devices and integrated circuits consisting of ZnO or derivative semiconductors.

ACKNOWLEDGEMENT

We would like to thank Mr. Ujihara, A., Dr. Ito, T. and co-researchers of Geomatec Co., Ltd. for preparation of the GZO films using a magnetron sputtering system, Mr. Maruyama, T. and co-researchers at Mitsubishi Gas Chemical Co., Inc. for preparation of the chemicals for patterning the GZO films, and Mr. Hokari, H. of Ortus Technology Co., Ltd., for fabrication of the 3 inch LCD panels. We are also grateful to Mr. Morisawa, K. and Mr. Osone, S. of Kochi university of Technology for their support, which aided in the development of the technique for froming the fine-patterns of the ZnO transparent films.

This work was made possible by grants for the Development of Indium Substitute Materials for a Transparent Conducting Electrode in the Rare Metal Substituted Materials Development Project from the Ministry of Economy, Trade and Industry and the New Energy and Industrial Technology Development Organization of Japan.

REFERENCES

1. Dong, B.-Z.; Fang, G.-J.; Wang,, J.-F.; Guan, W.-J.; & Xing-Zhong Zhao (2007), Effect of thickness on structural, electrical, and optical properties of ZnO: Al films deposited by pulsed laser deposition, *Journal of Applied Physics*, Vol. 101, 033713-1 - 033713-7, ISSN: 0021-8979.

2. Homma, S.; Miyamoto, A.; Sakamoto, S.; Kishi, K.; Motoi, N. & Yoshimura, K. (2005), Pulmonary fibrosis in an individual occupationally exposed to inhaled indium-tin oxide, Europian *Recipiratory Journal*, Vol. 25, No. 1, pp. 200–204, ISSN: 0903-1936.

3. Jin, Z. C.; Hamberg, I. & Granqvist, C. G. (1988), Optical properties of sputter-deposited ZnO:Al thin films, *Journal of Applied Physics*, Vol. 64, No. 10, pp. 5117-5131,

4. Kempthorne, D. & Myers, M. D. (January 12, 2007), *Mineral Commodity Summaries 2007*, United States Government Printing Office, pp. 78-79. prepared by Carlin, J. F. Jr. U.S. Geological Survey, ISBN-10: 0160778956, ISBN-13: 978-0160778957. http://minerals.usgs.gov/minerals/pubs/mcs/2007/mcs2007.pdf

5. Moss, T. S., (1980), Theory of intensity dependence of refractive index, *Physica Status Solidi (b)*, Vol. 101, No. 2, pp. 555-561, ISSN: 0370-1972

6. Nakagawara, O.; Kishimoto, Y.; Seto, H.; Koshido, Y.; Yoshino, Y.; Makino, T. (2009), Moisture-resistant ZnO transparent conductive films

with Ga heavy doping, *Applied Physics Letters*, Vol.89, No.9, pp. 091904 - 091904-3, ISSN: 0003-6951.

7. Ofuji, M.; Inaba, K.; Omote, K.; Hoshi, H.; Takanishi, Y.; Ishikawa, K. & Takezoe, H. (2002), Grazing incidence in-plane X-ray diffraction study on oriented copper phthalocyanine thin films", Japanese Journal of Applied Physics, Part 1, vol. 41, No. 8, pp. 5467–5471, ISSN: 0021- 4922.

8. Shin, S. H.; Shin, J. H.; Park, K. J.; Ishida, O. & Kim, H. H. (1999), Low resistivity indium tin oxide films deposited by unbalanced DC magnetron sputtering, *Thin Solid Films*, Vol. 341, issues 1-2, pp. 225 – 229, ISSN: 0040-6090.

9. Thornton, J. A. & Hoffman, D. W. (1977), Internal stresses in titanium, nickel, molybdenum, and tantalum films deposited by cylindrical magnetron sputtering, *Journal of vacuum science and technology*, Vol. 14, pp. 164- 168, ISSN 0022-5355.

10. Wakeham, S. J.; Thwaites,M. J.; Holton, B. W.; Tsakonas, C.; Cranton, W. M; Koutsogeorgis, D. C. & R. Ranson (2009), Low temperature remote plasma sputtering of indium tin oxide for flexible display applications, *Thin Solid Films*

11. Wortman J. J. & Evans, R. A. (1965), Young's Modulus, Shear Modulus, and Poisson's Ratio in Silicon and Germanium, *Japanese Journal of Applied Physics*, Vol. 36, pp. 153 – 156, doi:10.1063/1.1713863, ISSN: 0021-8979

12. Yamada, T.; Miyake, A.; Kishimoto, S.; Makino, H.; Yamamoto, N. & Yamamoto, T. (2007) Low resistivity Ga-doped ZnO thin films of less than 100 nm thickness prepared by ion plating with direct current arc discharge, *Applied Physics Letters*, Vol. 91, 051915, doi:10.1063/1.2767213 (*3 pages*). ISSN : 0003-6951.

13. Yamamoto, N.; Kume, K.; Iwata, S.; Yagi, K. & Kobayashi, N. (1986), Fabrication of highly reliable tungsten gate MOS VLSI's, *Journal of the Electrochemical Society*, Vol. 133, No.2, pp. 401-407, ISSN: 0013-4651.

14. Yamamoto, N.; Yamada, T.; Miyake, A.; Makino, H.; Kishimoto, S. & Yamamoto, T. (2008), Reletionship between residual stress and crystallographic structure in Ga-doped ZnO film, *Journal of the Electrochemical Society*, Vol. 155, No. 9, pp. J221-J225, ISSN: 0013-4651.

15. Yamamoto, N.; Makino, H.; Yamada, T.; Hirashima, Y.; Iwaoka, H.; Ito, T.; Ujihara, A.; Hokari, H.; Morita, H. & Yamamotoa, T. (2010), Heat resistance of Ga-doped ZnO thin films for application as transparent electrodes in liquid crystal displays, *Journal of the Electrochemical Society*, Vol. 157, No. 2, pp. J13-J20, ISSN: 0013-4651.

16. Yamamoto, N.; Makino, H.; Osone, S.; Ujihara, A.; Ito, T.; Hokari, H.; Maruyama, T. & Yamamoto, T. (2011a), Development of Ga-doped ZnO transparent electrodes for liquid crystal display panels, , *Thin Solid Films*, doi: 10.1016/j.tsf.2011.04.067, Article in press, ISSN: 0040-6090.

17. Yamamoto, N.; Makino, H.; Sato, Y. & Yamamoto, T, (2011b), Controlled formation of ZnO fine-pattern transparent electrodes by wet-chemical etching, ECS Transactions, Vol. 35, Issue 8, pp. 165 – 172, ISSN: 1938-5862.

18. Yamamoto, N.; Makino, H.; Sato, Y. & Yamamoto, T (2011c), Development of Ga–doped ZnO Transparent Electrodes as Alternatives for ITO Electrodes in Liquid Crystal Displays, the *SID 2011 digest of technical papers*, pp. 1375-1378, ISSN: 0097-966X.

19. Yamamoto, T.; Yamada, T.; Miyake, A.; Makino, H. & Yamamoto, N. (2008), Ga-doped zinc oxide: An attractive potential substitute for ITO, large-area coating, and control of electrical and optical properties on glass and polymer substrates, *Journal of the Society for Information Display*, Vol. 16/7, pp. 713-719, ISSN: 1071-0922.

CHAPTER 12

Low-voltage blue phase liquid crystal displays

Linghui Rao[a*] & Shin-Tson Wu[b]

[a] Microsoft Corporation, Redmond, WA, USA

[b] CREOL, College of Optics and Photonics, University of Central Florida, Orlando, FL, USA

ABSTRACT

Polymer-stabilised blue phase liquid crystal displays (PS-BPLCD) based on Kerr effect have become an increasingly important technology for information display applications. In comparison with conventional nematic LC devices, BPLCs exhibit several attractive features, such as sub-millisecond grey-to-grey response time, reasonably wide temperature range, no need for alignment layer, optically isotropic voltage-off state and large cell gap tolerance. However, some technical challenges such as high operation voltage, hysteresis, residual birefringence and relatively low transmittance remain to be overcome before their widespread applications can be realised. Recent progress on BPLC materials and devices has shown great promise. From material aspect, the electro-optical properties of blue phase liquid crystal material system are studied. To realise the electro-optic effect of PS-BPLC, novel device configurations that can dramatically improve the display performances are designed.

KEYWORDS

blue phase, liquid crystal displays, fast response time

1. INTRODUCTION

From cell phones, laptops, desktops, TVs to projectors, high-reliability liquid crystal displays (LCDs) have become indispensable in our daily life. Tremendous progress in LCDs has been made after decades of extensive research and development in materials, device configurations and manufacturing technology.[1]

Nowadays, the most critical issue on viewing angle has been solved using multi-domain structures [2–4] and optical film compensation.[5,6] Slow response time has been improved to 2–5 ms with low-viscosity LC material, overdrive and undershoot voltage, and thin cell gap approach. Moving image blur has been significantly reduced by impulse driving and frame insertion.[7] Contrast ratio in excess of 1 million-to-1 has been achieved through local dimming of the segmented LED backlight.[8–10] The colour gamut would exceed 100% of the NTSC (National Television System Committee), if RGB LEDs are used.[11] Besides these technological advances, the cost has been reduced dramatically by investing in advanced manufacturing technologies.

Polymer-stabilised blue phase liquid crystal displays (PS-BPLCDs) based on Kerr effect is emerging as a potential next-generation display technology. In comparison with conventional nematic devices, the PS-BPLCDs exhibit following attractive features: (1) sub-millisecond response time, (2) no need for molecular alignment layers, (3) optically isotropic dark state when sandwiched between crossed polarisers, and (4) transmittance is insensitive to cell gap when the in-plane electrodes are employed.[12,13] Therefore, BPLC holds promises for fast-response transmissive display, reflective display and photonics applications.

In this paper, we will review of the properties of PS-BPLC materials system and discuss the device operation mechanisms of LCDs based on the Kerr effect. New BPLC device structures are introduced to optimise display performances.

2. PROPERTIES OF PS-BPLC

2.1. PS-BPLC appearance and material system

Blue phase is a distinct thermodynamic phase that appears over a narrow temperature range (1–2°C) between the chiral-nematic/cholesteric (CH) and isotropic phases. Depending on the structure, there are up to three blue phases:

BP-I, BP-II and BP-III. BP-I and BP-II exhibit 3-D periodic structures in the director field as shown in Figure 1. They have body-centred and simple cubic symmetry, respectively. BP-III is seemingly amorphous; it has a same symmetry as the isotropic phase.[14]

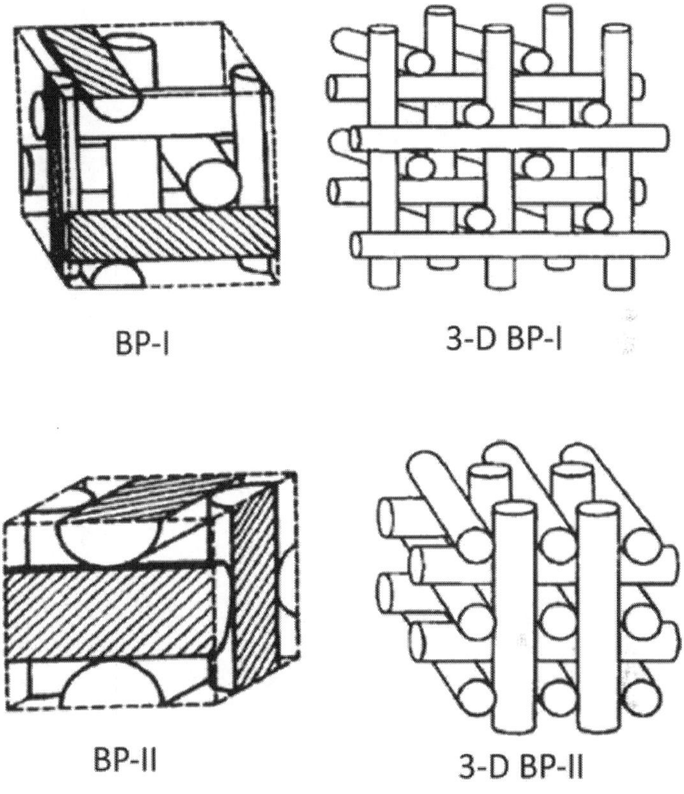

Figure 1. Cubic structures of BP-I and BP-II.

Figure 2 is the molecular structure of blue phase LC with cubic structure. The liquid crystal molecules have a double-twist alignment filled up in a cylinder. There is in fact unlimited number of helical axes presented; however, the structure was named double-twist. The LC directors perform a 45° rotation from the centre line. Due to the orientation of the LC molecules, microscopically, blue phase is optically anisotropic. The diameter of the cylinder is in the range the pitch length of the chiral nematic LC of ~100 nm. Because of this small range local reorientation, blue phase has a very fast response. So far, a microsecond-response BPLC with 39-μs decay time at room temperature has been demonstrated.[15] The double-twist cylinders further form the cubic symmetry; therefore, blue phase is optically isotropic macroscopically. The double-twist alignment is more stable than the single-twist structure such as the chiral nematics with normal helical structure. However, the double-twist cylinders cannot fill the cubic space completely uniformly to allow the directors to be matched everywhere. Meiboom has introduced the defect theory that defects occur at the intersections between the double-twist cylinders in the cubic structure.[16]The cubic structure of blue phase is stabilised by the lattice defects. The disclination lines in Figure 2 (d) are the defect lines.

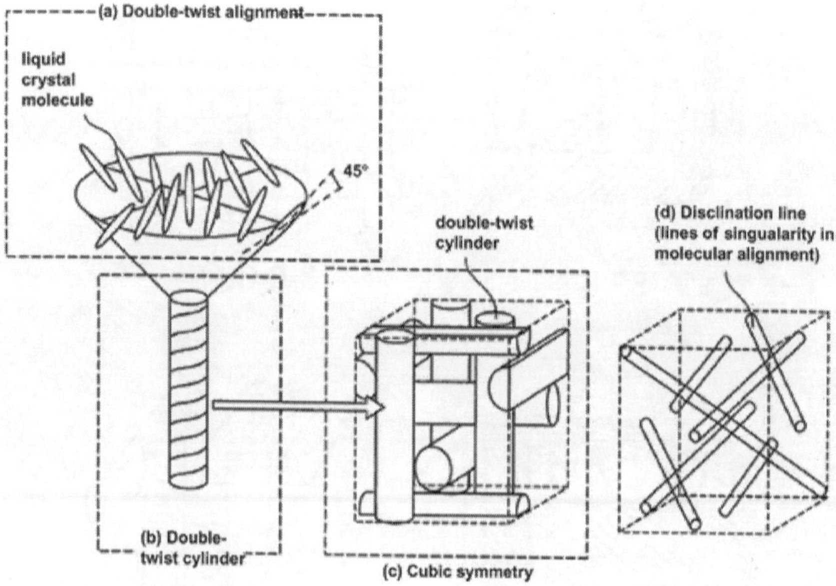

Figure 2. Blue phase LC structure at the microscopic level: (a) double-twist alignment of LC molecules; (b) double-twist cylinder; (c) lattice cubic symmetry formed by double-twist cylinders and (d) disclination lines of singularity in molecular alignment.

The temperature range in which the defect structure exists is very narrow (around 1–2°C) and is an issue problem from the viewpoint of practical applications. In 2002, Kikuchi proposed a polymer stabilisation method in which an equilibrium phase is thermodynamically stabilised by the coexistence of a polymer. Polymer chains are formed selectively in the disclination core, and therefore, the blue phase composite does not require a thermal energy to keep the disclination core isotropic at a temperature below isotropic phase. Through this method, the temperature range is reported to be expanded to more than 60° including room temperature.[17]

A PS-BPLC precursor typically comprises of 70–90 wt% of nematic LC host as switching molecules, 5–10 wt% of chiral dopant to induce blue phase, and 8–15 wt% of photocurable monomers and ~1% of photoinitiator for polymer stabilisation. To form BPLC composite, the precursor is cooled from isotropic phase to blue phase and then exposed to UV light. After UV irradiation, PS-BPLC nano-composites are self-assembled, exhibiting the characteristic small platelet structures.

2.2. Kerr effect

Kerr effect is a type of quadratic electro-optic effect caused by an electric-field-induced ordering of polar molecules in an optically isotropic medium. It usually exists in crystals with centro-symmetric point groups. If an electric field E is applied to a Kerr medium, such as PS-BPLC or optically isotropic LC composite, birefringence will be induced which is related to E as [18]:

$$\Delta n_{\text{ind}} = \lambda K E^2 \tag{1}$$

where λ is the wavelength and K is the Kerr constant. From Equation (1), the induced birefringence is linearly proportional to E^2.

The induced anisotropy will be along the direction of the electric field. Consequently, the isotropic sphere will appear as an elongated (Figure 3(b)) or a flattened (Figure 3(c)) ellipsoid, depending on whether the host LC has a positive or negative dielectric anisotropy ($\Delta\varepsilon$).[19] Equation (1) is valid only when the electric field is weak. As E keeps increasing, the induced birefringence will gradually saturate, as clearly described by the following extended Kerr effect [20]:

$$\Delta n_{\text{ind}} = \Delta n_s \left(1 - \exp\left[-(E/E_s)^2\right]\right) \tag{2}$$

where Δn_s denotes the saturated induced birefringence and E_s is the saturation electric field. In the weak field region ($E \ll E_s$), we can expand Equation (2) and deduce the Kerr constant as:

$$K \approx \Delta n_s / \left(\lambda E_s^2\right). \tag{3}$$

From Equation (3), a BPLC material with high Δn_s and low E_s will result in a large Kerr constant. Roughly speaking, Δn_s governs the optical property (e.g., maximum phase change) and E_s determines the electric property (operating voltage) of a BPLC material.

It has been reported by Gerber that the Kerr constant can be approximated by the following equation [21]:

$$K \sim \frac{\Delta n_{\text{ind}}}{\lambda E^2} \approx \Delta n \cdot \Delta\varepsilon \frac{\varepsilon_o P^2}{k\lambda(2\pi)^2}. \tag{4}$$

where Δn_{ind} is the induced birefringence, Δn, $\Delta\varepsilon$ and k are the intrinsic birefringence, dielectric anisotropy and elastic constant of the host LC material, respectively and P is the pitch length. Furthermore, we know that Δn, $\Delta\varepsilon$ and k are related to the nematic order parameter (S) as $\Delta n \sim \Delta n_o S$,[22] $\Delta\varepsilon \sim S/T$, and $k \sim S^2$.[23]

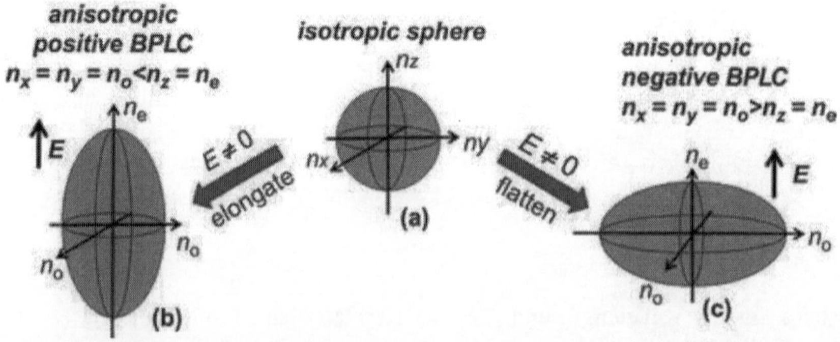

Figure 3. Electro-optic effect on BPLC refractive index ellipsoid: (a) BPLC without an electric field, (b) positive $\Delta\varepsilon$ BPLC with an electric field and (c) negative $\Delta\varepsilon$ BPLC with an electric field.

As is discussed in detail later, Kerr constant plays a key role on the operating voltage of the BPLC devices. A typical Kerr constant of BPLC composite is around 1 nm/V^2. In 2009, Kikuchi et al. [24] reported a polymer-stabilised isotropic phase composite with $K \sim 10$ nm/V^2 and its operating voltage in an IPS cell (5-μm electrode width and 10-μm electrode gap) is ~50 V_{rms}.

2.3. Response time

Fast response time is one of the most attractive features of PSBP LC since it helps to reduce motion blur and enable colour-sequential display while minimising colour break-up. Figure 4 shows the response time data from a paper by L. Rao about the prospects of PS-BPLCD.[25] The decay time is defined as 100–10% transmittance change.

As the temperature increases, the decay time decreases rapidly. At room temperature, the response time of BPLC is indeed experimentally shown being sub-millisecond. For example, we note from Figure 4 that the response time is decreased by 2X when the temperature is increased by merely 4–5°. This changing rate is ~3X more sensitive than that of most nematic LCs.

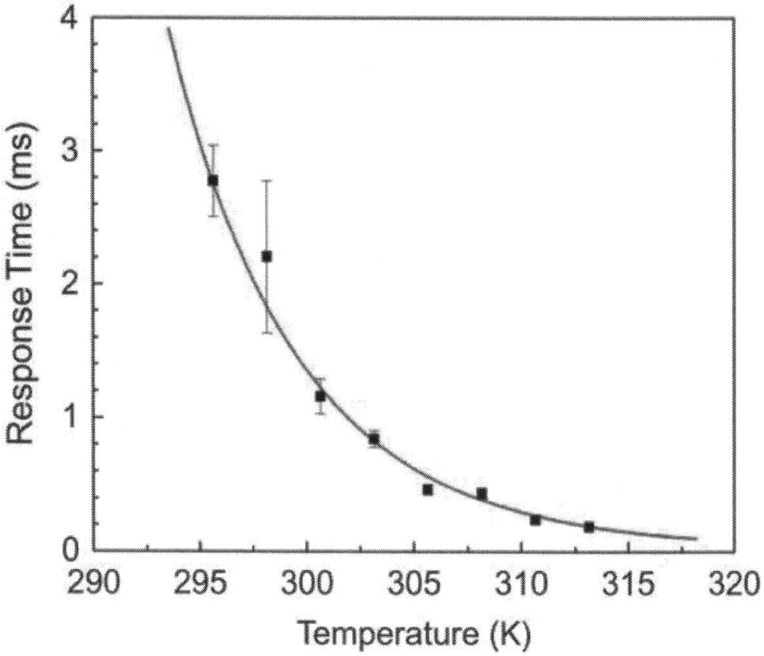

Figure 4. Temperature-dependent decay time. Dots are experimental data and line represents fitting using exponential curve in the paper.[25]

To understand this phenomenon, a physical model has been developed to explicitly correlate the temperature-dependent response time with LC parameters. The free relaxation time of a polymer-stabilised blue phase or optically isotropic LC composite can be approximated by [1,26]:

$$\tau \approx \frac{\gamma_1 P^2}{k(2\pi)^2}, \tag{5}$$

with γ_1 being the rotational viscosity, P the pitch length and k the elastic constant. The pitch length is insensitive to the temperature and can be treated as a constant.

From mean field theory,[23] elastic constant is related to the nematic order parameter (S) as $k \sim S^2$. With regard to rotational viscosity, the following modified Arrhenius model works well to describe the temperature-dependent rotational viscosity of nematic LCs [27,28]:

$$\gamma_1 \sim S \cdot \exp(E_a/K_B T), \tag{6}$$

where E_a is the activation energy of molecular rotation and K_B is the Boltzmann constant. With Equation (6) and treating P as a constant, we can rewrite Equation (5) as:

$$\tau \approx \frac{\gamma_1 P^2}{k(2\pi)^2} \sim \frac{S \cdot \exp(E_a/K_B T) \cdot P^2}{S^2 \cdot (2\pi)^2} \approx B \cdot \frac{\exp(E_a/K_B T)}{(1 - T/T_c)^\beta}. \quad (7)$$

From Equation (7), a low-viscosity host (smaller activation energy) is always in favour for achieving fast response time. However, there are trade-offs between response time and Kerr constant. For example, from Equation (4), a smaller elastic constant k and a larger pitch length P lead to a larger Kerr constant, but from Equation (7) the response time will be slower. Similarly, a larger dielectric anisotropy helps to enhance Kerr constant, but it could dramatically increase viscosity and thus slow down the response time. The polymer-stabilised blue phase or optically isotropic liquid crystal composite is a complex system, as it consists of nematic liquid crystal host, chiral agents and two types of monomers. Both material optimisation and UV curing process need to be optimised.

3. PS-BPLC FOR DISPLAYS

3.1. Blue phase liquid crystal displays with in-plane-switching fields

Typically, for display applications, the LC layer is sandwiched between two crossed linear polarisers. The operation mechanism of a PS-BPLCD is drastically different from conventional nematic LCDs in that the former is based on Kerr effect-induced isotropic-to-anisotropic transition while the latter relies on the anisotropic-to-anisotropic LC director reorientation. Moreover, the optic axis of the induced birefringence in BPLC is along the direction of the electric field. To achieve effective phase retardation, the induced birefringence should be parallel to the BPLC cell substrates, that is, a lateral electric field is desired. Among the present LCD modes, in-plane-switching (IPS) LCD can effectively generate horizontal electric fields; therefore, in blue phase LCD devices, IPS electrodes are commonly employed.[29]

Figure 5 illustrates the basic operation principles of a BPLCD in a planar IPS cell, where w is the electrode width and l is the electrode gap. When there's no electric field applied, the symmetric cubic structure in a PS-BPLC appears to be optically isotropic. Therefore, the linearly polarised light passing the bottom polariser experiences no phase retardation in the BPLC layer; thus, it cannot pass the crossed analyser. When an electric field is applied, the BPLC becomes anisotropic according to the Kerr effect; thus, the linearly polarised light passing the bottom polariser can pass the analyser on the top.

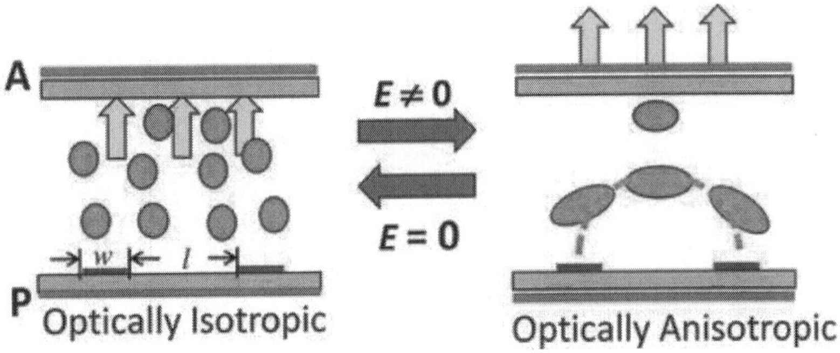

Figure 5. Operation principle of a BPLCD. Left: without electric field, BPLC appears to be optically isotropic; right: with electric field, BPLC becomes optically anisotropic.

The output transmittance of a uniaxial medium placed between two crossed linear polarisers is $T = T_0 \sin(2\phi)^2 \sin(\delta/2)^2$, where ϕ is the angle between optic axis of the uniaxial media and the absorption axis of the polariser, $\delta = 2\pi d\Delta n/\lambda$ is the phase retardation of the uniaxial medium. The angle ϕ needs to be 45° away from the linear polariser absorption axis in order to obtain maximum transmittance. In other words, the induced Δn also needs to be along that direction to have a maximum transmittance. Moreover, to get the maximum phase retardation δ, the induced birefringence should also be along the horizontal direction parallel to the substrate surface. Therefore, the stronger horizontal fields will benefit.

However, in a conventional IPS electrode configuration, the electric field is confined near the surface, and only the LC molecules in this shallow region contribute to the induced birefringence. As a result, the required voltage is high and transmittance is low. Even with a large Kerr constant of 13.7 nm/V^2 (Chisso JC-BP01M), the on-state voltage is still as high as ~50 V at room temperature, which is too high for amorphous-silicon thin-film transistors (a-Si TFTs).[30] Therefore, there is an urgent need to address this critical issue of high driving voltage.

3.2. BPLCDs with protrusion electrodes

To reduce the operating voltage, deeper penetration into the LC bulk region is needed. In the conventional planar IPS structure, we could strengthen the horizontal electric field by reducing the electrode gap. However, as the electrode gap becomes narrower, the penetration depth becomes shallower. The trade-off is the reduced optical efficiency. Figure 6 is proposed protruded electrode design with trapezoid shape.[31] The BPLC layer is interposed between two glass substrates, which are further sandwiched between two crossed polarisers; pixel and common electrodes are made with the trapezoid-shaped protrusion structure.

The trapezoid dimension is defined as follows: w_1 is the bottom width, w_2 is the top width, h is the height and l is the space between common and pixel electrodes. The protrusion structures have been widely used in multi-domain vertical alignment LCDs for obtaining wide viewing angle.[19] Besides, it can be made transparent with organic positive photoresist so the aperture ratio of the LC panel will not be sacrificed. In this design, the induced birefringence on top of the electrode does not contribute to the transmittance since the electric field is almost vertical. But compared with traditional IPS structure, stronger horizontal electric fields are generated between the pixel and common electrodes. Moreover, the field is able to penetrate deeply into the LC bulk region. The former plays a key role to reduce operating voltage while the latter helps to achieve high transmittance. This explains why a much lower on-state voltage is needed for the protruded electrode design.

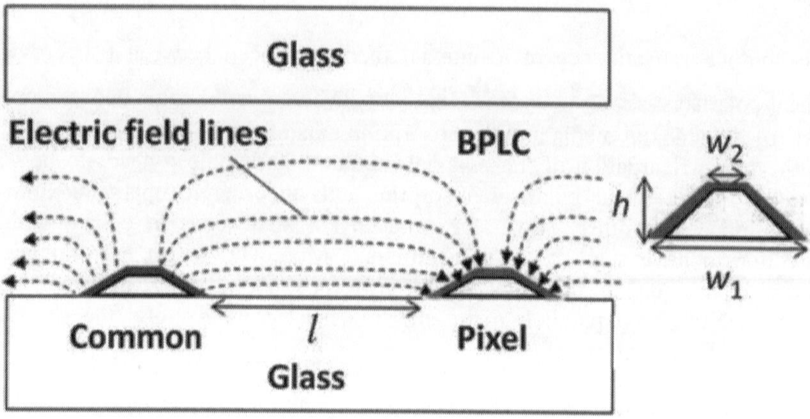

Figure 6. BPLCD structure with trapezoid-shaped protrusion electrodes and electric field lines.

For the protruded electrode with $w_1 = 2$ μm, $w_2 = 1$ μm, $h = 2$ μm and $l = 4$ μm shown in Figure 7, the peak transmittance is ~71% (normalised to the transmittance of two parallel polarisers) at 17 V_{rms}. Compared to the conventional IPS structure with electrode width $w = 2$ μm and electrode gap $l = 4$ μm (designated as IPS 2-4) with $V_{on} \sim 38$ V_{rms} and peak transmittance ~66.5%, our protrusion electrode shows a significant reduction in driving voltage while maintaining a relatively high transmittance.

To understand the underlying physics, we plot in Figure 8 the effective induced-birefringence (δn_{eff}) distribution for the protrusion electrode structure and IPS 2-4 structure. δn_{eff} is the effective birefringence that the normal incident light experiences. It is different from the induced birefringence Δn in Equation (1), which is the overall induced-birefringence of the BPLC material regardless of the electric field direction and the effectiveness whether the Δn will contribute to the transmittance.

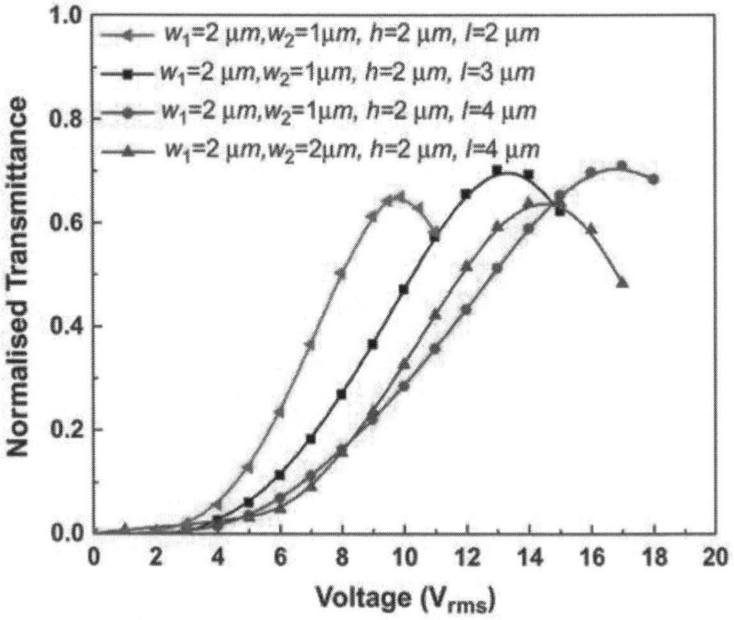

Figure 7. Normalised VT curves for protrusion electrodes with different dimensions.

From Figure 8, δn_{eff} is almost zero on top of the electrode for both structures since the electric field there is almost vertically distributed so that it does not provide phase change for the normal incident light. Nevertheless, the protrusion structure has a thicker penetrating depth and a larger δn_{eff} value in the effective region. This is the reason why a much lower on-state voltage is needed for the protrusion electrode design.

Figure 7 also shows the influence of electrode dimension on the voltage-dependent transmittance (VT) curve. Different electrode dimension generates different electric fields distribution in the BPLC cell. In comparison with the structure with parameters: $w_1 = 2$ μm, $w_2 = 1$ μm, $h = 2$ μm and $l = 4$ μm, if the spacing width l is decreased to 3 or 2 μm, then the on-state voltage is lowered to 13 or 9.9 V_{rms}, respectively. This is attributed to the stronger electric field generated as the electrode gap gets smaller. However, the trade-off is the lower transmittance. This is because the effective volume with horizontal electric field that accumulates the phase for the transmittance is smaller. If the size of w_2 is increased to 2 μm, the electric field near the bottom of the cell will be more uniformly distributed in horizontal direction. As a result, denoted by the black line in Figure 7, the operating voltage is reduced to 12 V_{rms}. Again, the trade-off will be that the transmittance is lower since larger w_2 reduces the effective spacing area between the electrodes. If we increase the protrusion height h to 4 μm, the peak transmittance would increase to 74.2% at 13 V_{rms}. The electric field in this case is stronger and more uniform as compared to the one with a 2-

μm height. Therefore, a lower driving voltage is needed. Meanwhile, the penetration of electric field gets deeper into the LC medium which enhances the transmittance. However, a 4-μm height protrusion electrode may be difficult to fabricate.

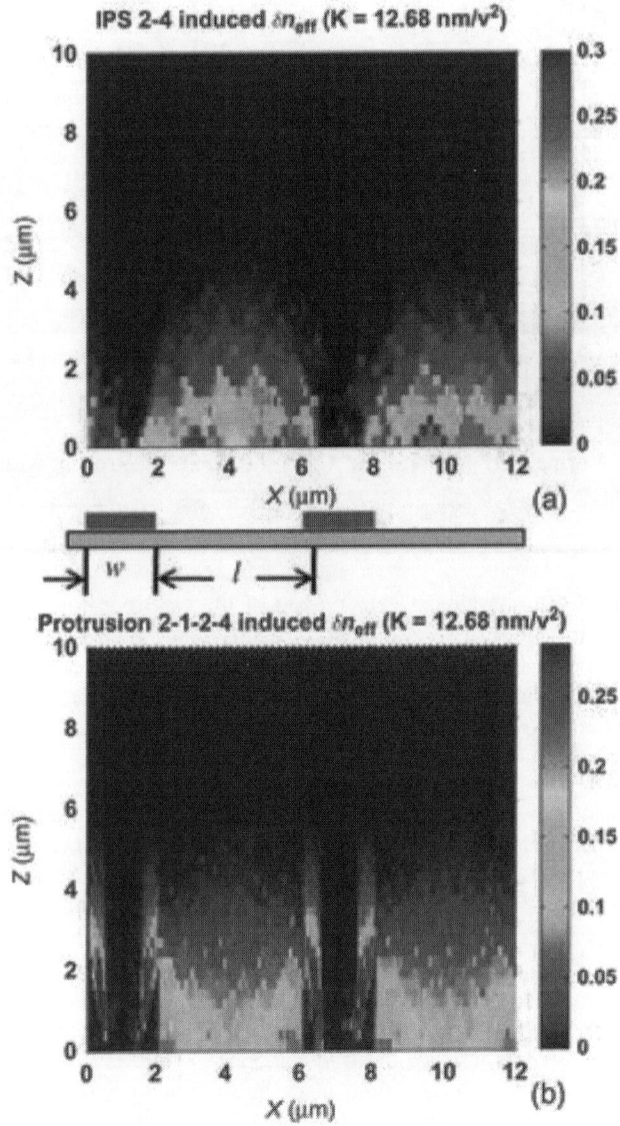

Figure 8. Induced effective birefringence (δn_{eff}) distribution for (a) conventional IPS cell with strip electrode with electrode width $w = 2$ μm, spacing $l = 4$ μm, and cell gap $d = 10$ μm, and (b) protrusion electrode with $w_1 = 2$ μm, $w_2 = 1$ μm, $h = 4$ μm and $l = 4$ μm, cell gap $d = 10$ μm.

In the voltage-off state, blue phase LC is optically isotropic so the dark state is very good. We have calculated the iso-contrast contours for the protrusion electrode structure with dimensions of $d = 10$ μm, $w_1 = 2$ μm, $w_2 = 1$ μm, $h = 2$ μm, $l = 4$ μm and $\lambda = 550$ nm. The contrast ratio (CR) of 200:1 is over 30° without any compensation film in Figure 9(a). In Figure 9(b), adding a half-wave biaxial film greatly suppresses the off-axis light leakage of the two crossed polarisers, so that CR = 1000:1 can reach about 60°. This result is even better than that of the strip-electrode IPS BPLC cell we reported previously because the peak transmittance for the protrusion electrode is slightly higher. While blue phase liquid crystals have a really good dark state, the CR is mainly governed by the on-state transmittance.

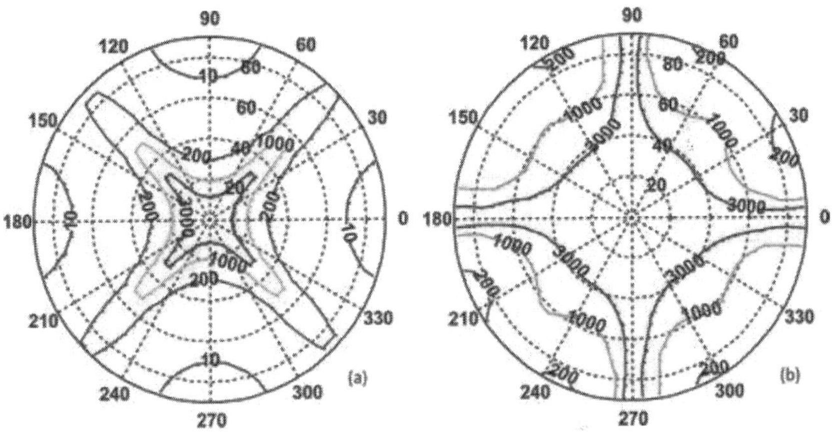

Figure 9. Iso-contrast plots of the BPLC cell with protrusion: (a) without compensation films and (b) with biaxial compensation films. The biaxial film parameters are: $N_z = 0.5$, $R_0 = (n_x - n_y).d = \lambda/2$.

At off-angle incident, the vertical part of the field is no longer vertical to the incident light, so the curved electric field will compensate the transmittance. Figure 10 shows the angular dependent (azimuthal and polar) light transmittance for (a) strip electrode (parameters: electrode width $w = 2$ μm and spacing width $l = 4$ μm) and (b) protrusion electrode (parameters: $w_1 = 2$ μm, $w_2 = 1$ μm, $h = 2$ μm and $l = 4$ μm). The polar angle of the protrusion structure with a transmittance of ~60% is over 50° while the strip electrode structure is only ~30°. The increasingly curved electric field in z-direction of protrusion electrode compensates more than the strip electrodes due to its deeper penetration layer into the LC bulk region. The iso-brightness contour plots in Figure 10 match the induced effective birefringence distribution in Figure 8.

To illustrate the importance of device structure, we compared the planar IPS electrodes with protrusion electrodes. From a previous experiment described in paper,[25] we have employed an IPS cell with a cell gap of 7.5 μm, and the ITO (indium tin oxide) electrode width of 10 μm and the electrode gap of 10 μm. We filled the same type of IPS cell with both the UCF BPLC material and Chisso BPLC material JC-BP01M. We find the device constant A is around 6 $\mu m^{1/2}$. Here, compared to the planar IPS cell, the protrusion electrode reported in Figure 11 has a device proportionality constant $A = 1.91\ \mu m^{1/2}$, which indeed indicates that a small A coefficient helps to lower the operating voltage.

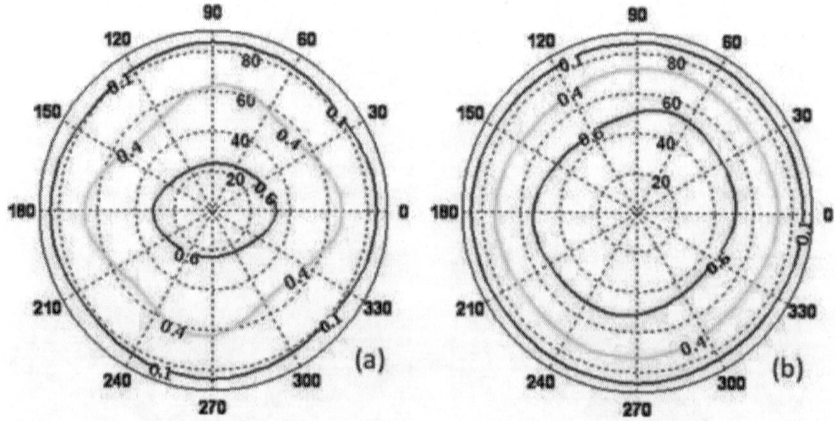

Figure 10. Angular-dependent (azimuthal and polar) light transmittance for: (a) strip electrode (parameters: electrode width $w = 2$ μm and spacing $l = 4$ μm) and (b) protrusion electrode (parameters: $w_1 = 2$ μm, $w_2 = 1$ μm, $h = 2$ μm and $l = 4$ μm).

In addition to optimising the device structure for generating strong and deep electric fields, it is equally important to develop LC materials with a larger Kerr constant. Figure 11 depicts the simulated VT curves at $\lambda = 550$ nm under different Kerr constants ($K_1 = 12.68$ nm/V^2, $K_2 = 3K_1$, $K_3 = 5K_1$) for the protrusion electrode structure with dimensions of $d = 10$ μm, $w_1 = 2$ μm, $w_2 = 1$ μm, $h = 2$ μm and $l = 4$ μm. With the increased Kerr constant, the on-state voltage is reduced from 17 to 9.7 V_{rms} and to 7.5 V_{rms}. The low-voltage operation would reduce the power consumption of the display devices. Amazingly, the on-state voltage is linearly proportional to$K^{-1/2}$, although the electric field generated from the protrusion electrodes is not uniform. Shown in the insert of Figure 11, the red line is the linear fitting for K_1, K_2, K_3, $K_4 = 2K_1$, and $K_5 = 4.5K_1$ with the linear equation as [30]:

$$V_{on} = A \cdot \frac{1}{\sqrt{K}}. \tag{8}$$

Here, the device proportionality constant $A = 1.91\ \mu m^{1/2}$. We can conclude that for a given structure, while Kerr constant K represents the influence from the material side to the driving voltage, the device constant A represents the effect from the device design. Larger Kerr constant K and smaller device proportionality constant A will lead to a lower operating voltage.

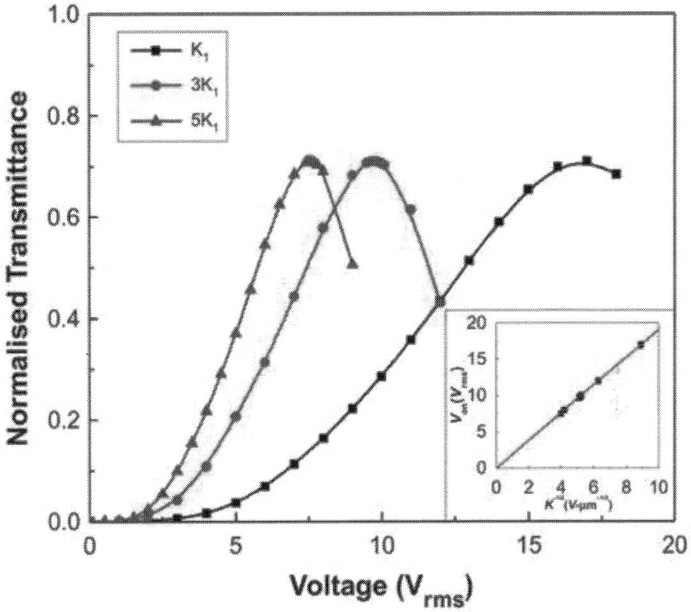

Figure 11. VT curves with different Kerr constants for protrusion electrode structure with $w_1 = 2$ μm, $w_2 = 1$ μm, $h = 4$ μm and $l = 4$ μm, cell gap $d = 10$ μm, inset is the linear plot between on-state voltage and $1/\sqrt{K}$.

The protrusion electrode structure is a feasible configuration for dramatically reducing the operating voltage of blue phase LCDs while keeping a reasonably high transmittance. There are already prototypes manufactured with this type of electrodes from leading LCD manufacturers such as Samsung and SEL lab.[32,33] By optimising the protrusion structure, the operating voltage can be successfully reduced to ~10 V_{rms}. This is an important milestone because the device can be driven by a-Si TFTs. There are other effects in optimising the device configuration, structures like BPLCDs with double-penetrating fringing fields,[34] with corrugated electrodes [35] and with vertical field switching [36] are proposed. With continuous development of large Kerr constant BPLC materials, the voltage can be reduced further.

4. REMAINING CHALLENGES

Although polymer-stabilised isotropic-to-anisotropic LCDs hold great potentials for future display applications, some technical challenges remain to be overcome:

1) Low operating voltage: The protrusion electrodes substantially lower the operating voltage, but its fabrication complexity is increased. Simpler methods for reducing voltage while keeping high transmittance need to be developed.

2) New materials: PS-BPLC represents a paradigm shift in materials. The conventional TFT-grade nematic LCs require a modest birefringence ($\Delta n \sim 0.1$) and dielectric anisotropy ($\Delta\varepsilon \sim 5$–10), low viscosity and high resistivity. However, to achieve a large Kerr constant ($K \sim 10$ nm/V^2), the BPLC would need to have a high Δn (>0.2) and large $\Delta\varepsilon$ (>30). These compounds will have long molecular conjugation and large dipole moment. As a result, their viscosity will be high. Moreover, the PS-BPLC requires two types of monomers and photoinitiator. The ionic impurities in these monomers would degrade the voltage-holding ratio. The temperature range is another concern. For display applications, the LC temperature range should cover from −40 to 80°C. In PS-BPLC, we could choose a wide-temperature-range nematic LC host. However, after mixing with chiral dopant, the mixture's clearing point often decreases substantially. For example, the clearing temperature of our LC host is 94°C, but it drops to 54°C after polymerisation. A low-clearing point will make the electro-optic properties of the LC/polymer composite very sensitive to the local temperature variation of a display panel, such as VT curves and response time. Consequently, the requirement for better thermal uniformity of a backlight unit is increased. A lot of new material development work (LC and monomers) along with the photopolymerisation process need to be studied.

3) Hysteresis: It is known that polymer-stabilised LCs often exhibit memory effect, also known as hysteresis. Some preliminary work on the hysteresis of BPLC and PS-BPLC has been reported [37]. Hysteresis affects the accuracy of grey-scale control and should be minimised.

4) Long-term stability: The concerns of long-term stability include twofold: incomplete photopolymerisation process and device operating lifetime, such as image sticking and polymer stability.

5) Residual birefringence: During the polymerisation process, some LC molecules could be embedded in the polymer networks. If this residual birefringence is not uniformly distributed in the entire panel, it will be difficult to be compensated by an optical film. As a result, the display CR will be degraded.

5. SUMMARY

PS-BPLC is emerging as a promising candidate for next-generation display and photonics devices. In this paper, we have reviewed the material system and electro-optic properties of PS-BPLC. The device optimisation for BPLCD is introduced that can successfully reduce the operating voltage to ~10 V. This is an important milestone as it enables the a-Si TFTs driving. With rapid advances in PS-BPLC materials and device structures, the widespread application of PS-BPLCs is foreseeable.

REFERENCES

1. Yang DK, Wu ST. Fundamentals of liquid crystal devices. New York (NY): Wiley; 2006.
2. Takeda A, Kataoka S, Sasaki T, Chida H, Tsuda H, Ohmuro K, Sasabayashi T, Koike Y, Okamoto K. 41.1: A super-high image quality multi-domain vertical alignment LCD by new rubbing-less technology. SID Symp Dig. 1998;29:1077–1080.
3. Oh-e M, Kondo K. Electro-optical characteristics and switching behavior of the in-plane switching mode. Appl Phys Lett. 1995;67:3895–3897.
4. Kim SG, Kim SM, Kim YS, Lee HK, Lee SH, Lee GD, Lyu JJ, Kim KH. Stabilization of the liquid crystal director in the patterned vertical alignment mode through formation of pretilt angle by reactive mesogen. Appl Phys Lett. 2007;90:2619110.
5. Hong Q, Wu TX, Zhu X, Lu R, Wu S-T. Extraordinarily high-contrast and wide-view liquid-crystal displays. Appl Phys Lett. 2005;86:121107.
6. Zhu X, Ge Z, Wu S-T. Analytical solutions for uniaxial-film-compensated wide-view liquid crystal displays. J Display Technol. 2006;2:2–20.
7. Hong H, Shin H, Chung I. In-plane switching technology for liquid crystal display television. J Display Technol. 2007;3:361–370.
8. Kim S. The world's largest (82-in.) TFT-LCD. SID Symp Dig. 2005;36:1842–1847.
9. Liu CT. Revolution of the TFT LCD technology. J Display Technol. 2007;3:342–350.
10. Lu R, Wu S-T, Lee SH. Reducing the color shift of a multi-domain vertical alignment liquid crystal display using dual threshold voltages. Appl Phys Lett. 2008;92:051114.

11. Lu R, Hong QI, Ge Z, Wu S-T. Color shift reduction of a multi-domain IPS-LCD using RGB-LED backlight. Opt Express. 2006;14:6243–6252.

12. Ge Z, Gauza S, Jiao M, Xianyu H, Wu S-T. Electro-optics of polymer-stabilized blue phase liquid crystal displays. Appl Phys Lett. 2009;94:101104.

13. Ge Z, Rao L, Gauza S, Wu ST. Modeling of blue phase liquid crystal displays. IEEE/OSA J Display Technol. 2009;5:250–256.

14. Crooker PP. Blue phases. In Kitzerow HS, Bahr C, editors. Chirality in liquid crystals. New York (NY): Springer; 2001.

15. Chen Y, Yan J, Sun J, Wu S-T, Liang X, Liu S-H, Hsieh P-J, Cheng K-L, Shiu J-W. A microsecond-response polymer-stabilized blue phase liquid crystal. Appl Phys Lett. 2011;99:201105.

16. Meiboom S, Anderson PW, Brinkman WF. Theory of the blue phase of cholesteric liquid-crystals. Phys Rev Lett. 1981;46:1216–1219.

17. Kikuchi H, Yokota M, Hisakado Y, Yang H, Kajiyama T. Polymer-stabilized liquid crystal blue phases. Nat Mater. 2002;1:64–68.

18. Kerr J. A new relation between electricity and light: dielectrified media birefringent. Philos Mag. 1875;50:337–348.

19. Rao L, Ge Z, Wu S-T. Viewing angle controllable displays with a blue-phase liquid crystal cell. Opt Express. 2010;18:3143–3148.

20. Yan J, Cheng H-C, Gauza S, Li Y, Jiao M, Rao L, Wu S-T. Extended Kerr effect of polymer-stabilized blue phase liquid crystals. Appl Phys Lett. 2010;96:071105.

21. Gerber PR. Electro-optical effects of a small-pitch blue-phase system. Mol Cryst Liq Cryst. 1985;116:197–206.

22. Wu S-T. Birefringence dispersions of liquid-crystals. Phys Rev. 1986;33:1270–1274.

23. Maier W, Saupe A. Eine einfache molecular-statistische theorie der nematischen kristallinflüssigen phase. Z Naturforsch Teil A. 1960;15:287–292.

24. Kikuchi H, Haseba Y, Yamamoto S-I, Iwata T, Higuchi H. 39.1: invited Paper: optically isotropic nano-structured liquid crystal composites for display applications. SID Int Symp Digest Tech Papers. 2009;40:578–581.

25. Rao L, Yan J, Wu S-T. Prospects of emerging polymer-stabilized blue-phase liquid-crystal displays. J Soc Inf Disp. 2010;18:954–959.

26. Gleeson HF, Coles HJ. Dynamic properties of blue-phase mixtures. Liq Cryst. 1989;5:917–926.

27. Wu S-T, Wu C-S. Experimental confirmation of the osipov-terentjev theory on the viscosity of nematic liquid-crystals. Phys Rev. 1990;42:2219–2227.

28. Wu S-T, Wu C-S. Rotational viscosity of nematic liquid crystals: a critical examination of existing models. Liq Cryst. 1990;8:171–182.

29. Soref RA. Transverse field effects in nematic liquid crystals. Appl Phys Lett. 1973;22:165–166.

30. Rao L, Yan J, Wu S-T, Yamamoto S-I, Haseba Y. A large Kerr constant polymer-stabilized blue phase liquid crystal. Appl Phys Lett. 2011;98:081109.

31. Rao L, Ge Z, Wu S-T, Lee SH. Low voltage blue-phase liquid crystal displays. Appl Phys Lett. 2009;95:231101.

32. Kubota D, Ishitani T, Yamashita A, Yamagata S, Oe Y, Tamura T, Ikenaga M, Yamamoto T, Kato M, Nakano M, Hatsumi R, Kubota Y, Murakawa T, Hayakawa M, Nishi T, Seo S, Hirakata Y, Yamazaki S, Okazaki K, Sato R, Cho T, Sakakura M. A new process for manufacture of low voltage polymer-stabilized blue phase LCDs. SID Int Symp Digest Tech Papers. 2011;42:125–128.

33. Lee H, Park H-J, Kwon O-J, Yun SJ, Park JH, Hong S, Shin S-T. The world's first blue phase liquid crystal display. SID Int Symp Digest Tech Papers. 2011;42:121–124.

34. Rao L, Cheng H-C, Wu S-T. Low voltage blue-phase LCDs with double-penetrating fringe fields. IEEE/OSA J Display Technol. 2010;6:287–289.

35. Jiao M, Li Y, Wu S-T. Low voltage and high transmittance blue-phase liquid crystal displays with corrugated electrodes. Appl Phys Lett. 2010;96:011102.

36. Cheng H-C, Yan J, Ishinabe T, Wu S-T. Vertical field switching for blue-phase liquid crystal devices. Appl Phys Lett. 2011;98:261102.

37. Rao L, Yan J, Wu S-T, Lai Y-C, Chen HY, Liang CC, Wu CM, Hsieh PJ, Liu SH, Cheng KL, Cheng K-L. Critical field for a hysteresis-free BPLC device. IEEE/OSA J Display Technol. 2011;7:627–629.

Index